KNOTTED DOUGHNUTS

AND OTHER MATHEMATICAL ENTERTAINMENTS

MARTIN GARDNER

KNOTTED DOUGHNUTS AND OTHER MATHEMATICAL ENTERTAINMENTS

W. H. Freeman and Company
New York

Library of Congress Cataloging-in-Publication Data

Gardner, Martin, 1914–
  Knotted doughnuts and other mathematical entertainments.

  Includes bibliographies and index.
  1. Mathematical recreations.   I. Title.
QA95.G27   1986   793.7′4   85-31134
ISBN 0-7167-1794-8
ISBN 0-7167-1799-9 (pbk.)

Printed in the United States of America

1 2 3 4 5 6 7 8 9 0   MP   4 3 2 1 0 8 9 8 7 6

To Gerry Piel and Dennis Flanagan

and all my other good friends
at *Scientific American*
during the 25 years
that I had the great privilege of writing
the magazine's Mathematical Games column.

Contents

Preface

Because this is the eleventh collection of my *Scientific American* columns, there is little to say in a preface that I have not said before. As in earlier volumes, I have made corrections and additions throughout and included addendums to provide material sent by readers and to update chapters in ways that were not easy to squeeze into the earlier text. References cited in the chapters are given more fully in the bibliographies that follow the chapters.

Martin Gardner

KNOTTED DOUGHNUTS

AND OTHER MATHEMATICAL ENTERTAINMENTS

CHAPTER ONE

Coincidence

Don't worry. Lightning never strikes twice in the same

— BILLY BEE

Since the beginning of history, unusual coincidences have strengthened belief in the influence on life of occult forces. Events that seemed to miraculously violate the laws of probability were attributed to the will of gods or devils, God or Satan, or at the very least to mysterious laws unknown to science and mathematics.

On the other hand, skeptics have argued that in the unthinkably intricate snarls of human history, with billions on billions of events unfolding every second around the globe, the situation is really the other way around. It is surprising that *more* strange coincidences are not publicized. "Life," wrote G. K. Chesterton in *Alarms and Discursions,* "is full of a ceaseless shower of small coincidences. . . . It is this that lends a frightful plausibility to all false doctrines and evil fads. There are always such props of accidental arguments upon anything. If I said suddenly that historical truth is generally told by red-haired men, I have no doubt that ten minutes' reflection (in which I decline to indulge) would provide me with a handsome list of instances in support of it."

"We trip over these trivial repetitions and exactitudes at every turn," Chesterton continued, "only they are too trivial even for conversation. A man named Williams did walk into a strange house and murder a man named Williamson. . . . A journalist of my acquaintance did move quite unconsciously from a place called Overstrand to a place called Overroads."

In his *Poetics,* Aristotle attributes to Agathon the remark that it is probable that the improbable will sometimes happen. All the same, most coincidences surely go unrecognized. For instance, would you notice it if the license plate of a car just ahead of you bore digits that, read backward, gave your telephone number? Who except a numerologist or logophile would see the letters U, S, A symmetrically placed in LOUISIANA or at the end of JOHN PHILIP SOUSA, the name of the composer of our greatest patriotic marches? It takes an odd sort of mind to discover that Newton was born the same year that Galileo died, or that Bobby Fischer was born under the sign of Pisces (the Fish). That's not all. "Fish" is chess slang for a mediocre player. In 1972, when Bobby Fischer's blunder cost him the first game in his famous match in Iceland with Boris Spassky, he said afterward, "I'm a fish! I played like a fish!"

There are two other reasons why strange coincidences are seldom recorded. When trivial ones are noticed, it is easy to forget them, and when they are remarkable enough to be remembered, one may hesitate to speak about them for fear of being thought superstitious. Skeptics maintain that with all of this in mind the number of astonishing coincidences that continually occur as the result of ordinary statistical laws is far greater than even occultists realize.

The ancient view that many coincidences are too improbable to be explained by known laws has recently been revived by Arthur Koestler. In his book *The Roots of Coincidence,* he devotes many pages to a theory developed by Paul Kammerer, an eccentric Austrian biologist, whose Lamarckian convictions were much admired by T. D. Lysenko and who was the hero of Koestler's previous book, *The Case of the Midwife Toad.* Kammerer wrote a book, *Das Gesetz der Serie* (1919), about his theory of coincidences. It describes exactly 100 coincidences — concerning words, numbers, people, dreams and so on — that he had collected over a period of 20 years.

Kammerer's seventh coincidence is typical. On September 18, 1916, his wife was in a doctor's waiting room admiring magazine reproductions of paintings by a man named Schwalbach. A door opened and the receptionist asked if Frau Schwalbach was in the room. Kammerer's 10th coincidence is even more impressive. Two soldiers were separately admitted to the same hospital. They were 19, had pneumonia, were born in Silesia, were volunteers in the Transport Corps and were named Franz Richter.

Kammerer was persuaded that such oddities could be accounted for only by assuming a universal law, independent of physical causality, that brought "like and like together." Koestler is sympathetic to this view. He suggests that some of the results of parapsychology, such as the tendency of falling dice to show a certain number more often than expected, can be explained not as the influence of mind on matter but as coincidences produced by a transcendent "integrative tendency."

Estimating the probability that a hidden law is at work behind a series of apparent coincidences is a difficult task, and statisticians have developed sophisticated techniques for doing so. How easy it is for our intuitions to go astray is illustrated by many familiar paradoxes. If 23 students are in a classroom and you pick two at random, the probability that their birthdays (month and day) match is about 1/365. The probability that at least two of the 23 have the same birth date, however, is a trifle better than 1/2. The reason is that now there are $1 + 2 + 3 + . . . + 22 = 253$ possible matching pairs, and figuring the exact probability of coincidence is a bit tricky.

In a class of 35 students the probability of a birthday coincidence rises to about 85 percent. If students call out their birth dates one at a time until someone raises a hand to indicate that his birthday matches the one just called, you can expect a hand to go up after about nine calls (see "Note on the 'Birthday Problem,'" by Edmund A. Gehan in *The American Statistician*, April, 1968, page 28). William Moser has pointed out that the chances are better than even that two people in a group of 14 will have birth dates that either are identical or fall on consecutive days of the year. Among seven people, he calculates, the probability is about 60 percent that two will have birthdays within a week of each other, and among four people the probability is about 70 percent that two will have birthdays within 30 days of each other.

Variants of the basic idea are endless. The next time you are in a gathering of a dozen or more people try checking on such things as the exact amount of change each person has, the first names of his parents, the street numbers of his home, the playing card each writes secretly on a slip of paper and so on. The number of coincidences may be scary.

Another simple demonstration of an event that seems improbable but actually is not can be given with a deck of playing cards. Shuffle the cards, then deal them while you recite their names in a predetermined order, say ace to king of spades followed by the same sequence for hearts, clubs and diamonds. The probability that a card named in advance, such as the queen of hearts, will be dealt when it is named is 1/52, but the probability that at least one card will be dealt when named is almost 2/3. If you name only the values, the probability of a "hit" rises to 98 percent, or very close to certain.

In the foregoing instances the probabilities can be calculated precisely. For most events in daily life, however, probability estimates of coincidences are necessarily vague. For example, a great deal of research has been done on the "small-world problem." What is the probability that if you meet a stranger on an airplane, the two of you will have at least one acquaintance in common? Not only are accurate statistics hard to come by but also the very terms of the problem are impossible to define precisely. Who, for instance, is an "acquaintance"?

In spite of such formidable difficulties there is strong evidence that it is indeed a smaller world than most people imagine. Suppose a person is given a document and asked to transmit it to someone he does not know who lives in another city in another part of the U.S. The procedure is to send the document to a friend whom he knows on a first-name basis and who seems the most likely to know the "target" person. The friend in turn then sends the document to one of *his* friends with the same instructions, and the chain continues until the document reaches the target. How many intermediate links will the chain have? Most people guess about 100. When psychologist Stanley Milgram made actual tests, he found that the links varied from two to 10 and that the median was five.

Pick two women at random. The probability that both are wearing green shoes is low, but if you consider 20 ways the women can match — color of eyes, first names, type of hairdo and so on — the probability of a coincidence is close to certainty. It is hard to believe, but gross miscarriages of justice have resulted from a failure to understand just such trivial truths. In 1964 a black man and his white wife were convicted of a mugging in San Pedro, Calif., mainly because they were the only couple in the area who matched the reports of witnesses on five counts: the girl was a blonde, she had a ponytail, her companion was black, he had a beard, they drove a yellow car. The prosecutor estimated each probability separately — 1/10 for a yellow car, 1/1,000 that a couple are black and white, and so on — then he multiplied the five fractions and convinced the jury that the probability was 1/12,000,000 that a matching couple lived in the vicinity. Not until four years later (see *Time*, April 26, 1968, page 41) did the California Supreme Court reverse the decision after a judge less ignorant of mathematics persuaded the court that the estimate should have been about 41/100.

Anyone who watches carefully for coincidences involving himself can easily find them. "Did you ever notice that remarkable coincidence?" F. Scott Fitzgerald wrote in 1928 to the British writer Shane Leslie. "Bernard Shaw is 61 years old, H. G. Wells is 51, G. K. Chesterton is 41, you're 31, and I'm 21 — all the great authors of the world in arithmetical progression." Carl Sandburg was quoted in *The New York Times*, January 6, 1967, as saying that having completed his 89th birthday he confidently expected to live to 99. He had two great-grandfathers and a grandfather who had died in years that were multiples of 11. Having got safely past 88, Sandburg expected to go on to 99. Unfortunately he died six months later. Lewis Carroll recorded in his diary that most good things that happened to him, of which the best were meeting new and comely little girls, occurred on Tuesdays.

Surely the strangest coincidence involving a major U.S. magazine was the case of the "deadly double" ads in *The New Yorker,* November 22, 1941, which generated rumors about Japanese undercover agents for many years after. The long-submerged rumors surfaced in 1967 when a former U.S. naval intelligence agent, Ladislas Farago, told the story in a press release for his book *The Broken Seal,* an account of American and Japanese intelligence operations before World War II. Sixteen days before Pearl Harbor *The New Yorker* ran two advertisements (pages 32 and 86) for a new dice game called The Deadly Double *[see Figure 1].* Were these advertisements placed by the Japanese to inform their undercover agents of the planned attack on Pearl Harbor?

Farago's press release pointed out the following correlations. The attack was on December 7. In the smaller first advertisement, note the 12 (for December) on one die and the 7 on the other. Above the dice are the words "Achtung. Warning. Alerte!" The numbers 5 and 0, Farago said, could have indicated the planned time for the attack, which did not start until 7:00 A.M. The XX, or 20, is the approximate latitude of Pearl Harbor. Farago admitted that he did not know what the 24 stood for.

The second advertisement shows two people playing the dice game during an air raid, with the XX repeated on the symbol of the double-headed eagle. A *Times* story of March 12, 1967, based on Farago's press release, stated that the mysterious dice game had never existed. Farago told the *Times* that he had first learned of the ads from his friend Al Hirschfeld, the newspaper's theatrical caricaturist. When Farago questioned officials at *The New Yorker,* he said, "They were very closemouthed about it."

These fantastic allegations were quickly dissipated by the *Time's* follow-up story on March 14. The dice game *did* exist. Mrs. E. Shaw Cole, widow of the man who invented it, had been found in Montclair, N.J. She had helped her late husband, Roger Paul Craig, write the ads. Several New York department stores were selling the game in 1941. Agents of the Federal Bureau of Investigation, Mrs. Cole said, actually had visited them after the Pearl Harbor attack, but any relation between the attack and the ads was just "one big coincidence."

"What can I say?" said Farago.

Several years ago I asked Dr. Matrix, the famous numerologist, for his opinion on the advertisements. The XX, he told me, indicates that two *X*'s are to be appended to the alphabet. The first number on the die, 12, instructs us to count to the 12th letter, *L.* The second number 24, tells us to count 24 letters forward from *L,* including of course the extra *X*'s, and carrying the count back

Figure 1 Two advertisements that appeared in *The New Yorker* for November 22, 1941.

to the beginning. This second count ends on *H*. The 7 on the die at the right tells us to count seven letters forward from *H* to *O*. The three letters found in this straightforward manner are *L*, *H* and *O*, the initials of Lee Harvey Oswald. The advertisements in *The New Yorker* appeared in the November 22, 1941, issue. November 22 was the date of President John F. Kennedy's assassination, and 22 added to 1941 is 1963, the year of the assassination.

It is easy to understand how anyone personally involved in a remarkable coincidence will believe that occult forces are at work. You can hardly blame the winner of the Irish Sweepstakes for thinking that Providence has smiled on him even though he knows it is absolutely certain that *someone* will win. Gamblers are particularly susceptible to this belief, and they tend to be more superstitious than most. In every big city in the U.S. there are thousands of policy "hunch players" who like to bet on numbers prominent in the news. It is hardly surprising that now and then such hunches pay off. In 1958, for example, 48 people died when a Jersey Central commuter train plunged into Newark Bay. The last car taken from the water was shown in newspapers and on television with its number 932 clearly visible. Thousands of Manhattan policy players bet on 932 and won. A similar coincidence was reported in *The New York Times* for January 24, 1967. The President's daughter, Luci Johnson Nugent, had just given birth to a boy weighing eight pounds 10 ounces. All over Brooklyn bets were made on various permutations of these three digits. When 081 won, Brooklyn policy banks were closed for days because of the losses.

In science, as in daily life, it is not always easy to know if an observed correlation of "like and like" is pure coincidence or evidence of underlying structure. It was coincidence (plus some fudging) that the planetary orbits fitted Kepler's pattern of nested Platonic solids but not coincidence that data on their orbits fitted his patterns of ellipses. It is undoubtedly coincidental that the disks of the sun and moon, seen from the earth, are almost exactly the same size. The sun's diameter is 400 times that of the moon, but incredibly it is just 400 times as far away, as though nature planned it that way to give us a spectacular display of the sun's corona during a total eclipse. On the other hand, for half a century most geologists were convinced that the fit of the edges of the land masses on each side of the Atlantic was sheer coincidence. Alfred L. Wegener's theory that the two land masses had once been a supercontinent that had split and drifted apart (a notion that had been advanced by Francis Bacon) was considered crankish until about 10 years ago. Now it is the preferred hypothesis.

There are similar difficulties in mathematics. The curious repetition of 1828 in the first nine decimals of *e* (2.718281828 . . .) is almost certainly coincidental. Consider now the square roots of .999 and .9999999. They are respectively .9994 . . . and .99999994. . . . Is it accidental that in each case the

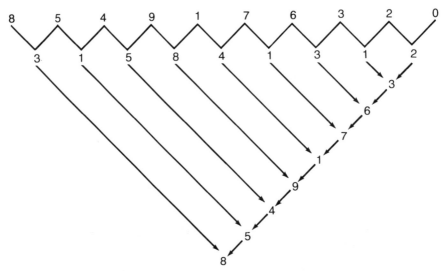

Figure 2   Benson Ho's answer.

irrational square root of a decimal fraction consisting of *n* 9's begins with *n* 9's followed by a 4? No, as Richard G. Gould has pointed out in a letter; it can be shown to be true of all such "rep-9" decimal fractions. You have only to express their square roots as $(1 - 10^{-a})^{1/2}$, expand the expression by the binomial theorem and interpret the results properly to establish the theorem.

The number 4 is a square number, and if you append to it the next consecutive square number, 9, the result is 49, another square. Is it a coincidence or a special case of a general law? One more curious question (both will be answered next month): An old brainteaser asks for the ordering principle behind the sequence 8549176320, which contains all 10 digits. The answer is that they are in the alphabetical order of their names. When the Massachusetts Institute of Technology's *Technology Review* printed this answer in its issue of July, 1967, page 10, it added a second answer that had been supplied by a reader named Benson P. Ho. His solution is best explained by his diagram *[see Figure 2]*. The digit above the right arm of each *V* is subtracted from the digit above the left arm. If the result is negative, add 10. The result goes under each *V*. Arrow pairs point to digits that are the sum of the two digits at the back of each arrow. If the sum is greater than 10, subtract 10. Note that the diagonal series of digits, when they are read upward, repeats the original series. It is a remarkable coincidence. Or is it?

ANSWERS

Neither of the two numerical oddities are coincidences.

S. N. Collings, in *The Mathematical Gazette* (December, 1971, page 418), generalizes the fact that joining consecutive squares 4 and 9 produces the square 49 as follows: Let $(n - 1)^2$ and n^2 be two consecutive squares. Join them to form a two-digit number in a notation with a base of $n^2 + 1$. (In the case of $2^2 + 3^2$ the base is $3^2 + 1 = 10$.) The new number will be $(n - 1)^2 \times (n^2 + 1) + n^2$, which equals the square number $(n^2 - n + 1)^2$.

Philip G. Smith, Jr., discovered that a reverse procedure always yields the same square. Interpret each of the squares in a base equal to the smaller square plus 1, put the larger of the two squares in front of the smaller and interpret the result in a base equal to the smaller square plus 1. In decimal notation: consecutive squares 9 and 16 join to produce square number 169. If the opposite procedure is followed, the result is 9 followed by 16, with 16 regarded as a single symbol of base-17 notation. The number's decimal equivalent is $(9 \times 17) + 16 = 169$, the same square that was obtained before.

On the surface it seems surprising that both procedures always give the same result, but, as Smith showed, it is merely a special case of the following general theorem. Let x and y be any positive real numbers. If both are expressed in base $x + 1$, and x is appended to y, the value is the same as expressing the numbers in base $y + 1$ and appending y to x. In the first case the value is $y(x + 1) + x$, and in the second $x(y + 1) + y$. The two expressions are clearly equivalent.

The pattern that Benson P. Ho found for the series 8549176320 is a ho, ho, ho hoax. It is not hard to show that any series of digits ending in 0, subjected to Ho's procedure, will give the same result.

ADDENDUM

Judith Bronowski wrote to correct my statement that Francis Bacon anticipated continental drift. It is true that in the second book of *Novum Organum* (Section 28), Bacon spoke of the remarkably similar shapes of the Atlantic coasts of South America and Africa, but his only explanation was that this could "not be attributed to mere accident." The earliest known record of explaining this seeming coincidence by assuming that a continent split and the two parts drifted from each other is, according to Bronowski, in a book called *The Creation and Its Mysteries Revealed,* by Antonio Snider-Pelligrini (Paris, 1858). The book had no influence on geologists. Wegener was apparently the first to suggest continental drift as a serious scientific theory.

BIBLIOGRAPHY

"Coincidences." William S. Walsh in *Handy-Book of Literary Curiosities,* pages 170–174. Lippincott, 1892.

"The Small World Problem." Stanley Milgram in *Psychology Today,* May, 1967, pages 61–67.

The Roots of Coincidence. Arthur Koestler. Random House, 1972. My review of this book is reprinted in my *Science: Good, Bad and Bogus,* Prometheus Books, 1981, Chapter 22, along with Koestler's rebuttal and my reply to Koestler. See also N. T. Gridgeman's more detailed review in *Philosophy Forum,* Vol. 14, 1975, pages 307–316.

The Challenge of Chance. Alister Hardy, Robert Harvie and Arthur Koestler. Random House, 1973.

Incredible Coincidences. Alan Vaughan. Lippincott, 1979. An outstanding instance of a naive book by an occultist with no comprehension of statistics or any awareness of the danger of taking anecdotes as scientific evidence.

"On Coincidence." Ruma Falk in *The Skeptical Inquirer,* Vol. 6, 1981–1982, pages 18–31.

"Mere Coincidence?" Robert A. Wilson in *Science Digest,* January, 1982, pages 84–85, 95.

"Against All Odds." Richard Blodgett in *Games,* November, 1983, pages 14–18.

"The Powers of Coincidence." Rudy Rucker in *Science 85,* February, 1985, pages 54–57.

The Magic Numbers of Dr. Matrix. Martin Gardner. Prometheus Books, 1985.

CHAPTER TWO

The Binary Gray Code

The binary Gray code is fun,
For in it strange things can be done.
  Fifteen, as you know,
  Is one, oh, oh, oh,
And ten is one, one, one and one.

    — ANON.

Although the decimal system is now in common use throughout the world, mathematicians and computers often manipulate integers by using other systems, some with such exotic features as mixed bases, negative bases, irrational bases or floating points. One of the most useful of these systems — one with surprising puzzle applications — is the Gray code.

The first puzzle application of a Gray code, which I shall describe below, was in 1872, when a binary version provided an elegant solution to a much older mechanical puzzle. The term "Gray," however, derives from Frank Gray, a research physicist at the Bell Telephone Laboratories, who died in 1969. His contributions to modern communication technology were immense. The method now in use for compatible color television broadcasting was developed by Gray (numerologists note!) in the 1930's. In the 1940's he devised what was soon to be called the binary Gray code to avoid the large errors that could arise in transmitting signals by pulse code modulation (PCM). The first publication of this code was in his U.S. Patent 2632058 (March 17, 1953) for a Gray coder tube that eliminated the quantizing grid wires used in early PCM transmission tubes.

Exactly what is a Gray code? It is a way of symbolizing the counting numbers in a positional notation so that when the numbers are in counting order, any adjacent pair will differ in their digits at one position only, and the absolute difference at that position will be 1. For instance, 193 and 183 could be adjacent counting numbers in a decimal Gray code (the middle digits differ by 1), but not 193 and 173, nor 134 and 143. There is an infinity of Gray codes, since they apply to any base system and for each base there are many different ways to construct the code.

To appreciate the value of such a system, consider what happens when the odometer of a car reads 9,999 miles. To register the next mile, five wheels must rotate to show 10,000. Because the wheels move slowly, there is little chance of error. But if counting is recorded electronically at enormously high speeds, when two or more digits change simultaneously the likelihood of producing a false number zooms upward. The probability is greatly reduced if the counting procedure requires only one decision whenever the magnitude to be coded is halfway between two adjacent quantized steps, regardless of whether the magnitude is increasing or decreasing. If the counting is by Gray code, only one digit of the counter changes by only one unit at each step.

The mileage meter is a familiar example of what are called analog-to-digital (A/D) converters. A continuous (in this case always increasing) variable, the mileage (or, if you prefer, the number of times the car wheels have rotated) is given a digital output. There are many other control systems in which analog-to-digital conversion must proceed at enormously high speed while the variable being measured fluctuates rapidly. Examples include wind-tunnel simulations of airplanes and guided missiles, and PCM applications where voltages, shaft positions, wave amplitudes of sounds, colors and so on must be translated almost instantly to a digital output signal. At one time a human observer would take pointer readings or inspect a curve on a graph, record the magnitude in digital form and feed this information to a computer. Today the slow and errorprone middleman is eliminated by analog-to-digital converters connected directly to the computer. A great increase in accuracy and often a considerable saving in hardware result from counting scales in Gray codes.

Binary Gray codes are the simplest. If we limit the code to one digit, there are only $2^1 = 2$ numbers, 0 and 1. Disregarding reversals, there is only one Gray code: 0, 1. We can graph this as a straight line, its ends labeled 0 and 1 [see Figure 3, left]. The Gray code is obtained by moving along the line in either direction. A Gray code for two binary digits has $2^2 = 4$ numbers: 00, 01, 10 and 11. The corners of a square can be labeled with these numbers [see Figure 3, middle]. The labeling is such that the binary numbers at any pair of adjacent corners differ in only one place. We can start at any corner and visit all four

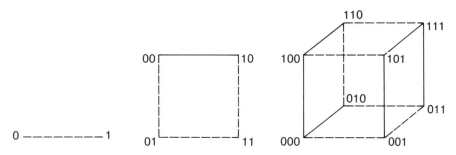

Figure 3   Graphs for one (left), two (middle) and three-digit (right) binary Gray codes.

corners by going clockwise or counterclockwise around the square. If we ignore reversals, this produces four Gray codes. The line starting at 00 yields the Gray code 00, 01, 11, 10. The code is cyclic because the path can return to 00.

A Gray code for three-digit binary numbers has $2^3 = 8$ numbers that can be placed on the corners of a cube *[see Figure 3, right]*. Adjacent corners have binary triplets that differ in only one place. Any continuous path that visits every corner once only generates a Gray code. For example, the path shown by the dashed line starting at 000 produces 000, 001, 011, 010, 110, 111, 101, 100. This is a cyclic code because the path can return from 100 to 000 in one step. Such paths are called Hamiltonian paths after the Irish mathematician William Rowan Hamilton. As the reader has probably guessed, binary Gray codes correspond to Hamiltonian paths on cubes of n dimensions. A Gray code for four-digit binary numbers has $2^4 = 16$ numbers that fit the corners of a hypercube in 4-space, for five digits a hypercube in 5-space and so on. Interested readers will find this covered in detail in E. N. Gilbert's paper (see the bibliography).

Gray codes for other bases correspond to Hamiltonian paths on more complicated n-dimensional graphs. The number of Gray codes for any base increases explosively as the number of digits increases. The number of Gray codes, even for the binary system, is known only for four or fewer digits.

An ill-fated attempt to obtain the number for five binary digits is recounted in *Graph Theory and Its Applications,* by Ronald C. Read, who wrote a BFI program for finding the number of Hamiltonian paths on the five-dimensional cube. BFI is Read's acronym for brute force and ignorance. ("It should be BFBI," he has since remarked, "the second B standing for 'Bloody,' but one has to preserve a measure of decorum in published papers.") "These are algorithms," he explains, "devoid of any subtlety whatever, which simply keep thumping the problem on the back until it disgorges an answer." After the program ran for a short time (on a computer in Kingston, Jamaica), a sample of the output was examined in order to estimate how long the run would be. The guess was 10 hours, and so the computer was set to run unattended overnight. During the night a tropical thunderstorm cut the power supply, and the computer stopped.

"Idle curiosity," Read continues, "prompted us to look to see where the program had got to before being so abruptly terminated, and in doing so we discovered that we had made a rather serious error in calculating our previous estimate of the running time. Our revised estimate turned out to be more like ten years!"

Read sensibly abandoned the project. The problem was not solved until 1980 (see the addendum).

For practical purposes it is important to select a Gray code with two desiderata: (1) rules for its formation should apply to the entire set of counting numbers; (2) it should have simple conversion rules for translating a standard number to its Gray code equivalent and vice versa.

The simplest Gray code with both features is called a reflected Gray code. For most mathematicians it is *the* Gray code. To convert a standard binary number to its reflected Gray equivalent, start with the digit at the right and consider each digit in turn. If the next digit to the left is even (0), let the former digit stand. If the next digit to the left is odd (1), change the former digit. (The digit at the extreme left is assumed to have a 0 on its left and therefore remains unchanged.) For example, applying this procedure to binary number 110111 gives the Gray number 101100.

To convert back again, consider each digit in turn starting at the right. If the sum of all digits to the left is even, let the digit stay as it is. If the sum is odd, change the digit. Applying this procedure to 101100 restores the original binary number 110111.

Inspection of the numbers from 0 through 42 and their reflected binary Gray code equivalents will show that every two adjacent Gray numbers differ at only one place, and of course the difference is necessarily 1 [see Figure 4]. It is called a reflected code because the series can be generated rapidly by the following algorithm. Start with 0, 1 as a one-digit Gray code, then reflect (reverse) and append the digits to get 0, 1, 1, 0. Next put 0's in front of the first two numbers and 1's in front of the last two numbers. The result is a two-digit Gray code: 00, 01, 11, 10. To extend the series to three-digit Gray numbers, reflect the two-digit code 00, 01, 11, 10, 10, 11, 01, 00. As before, put 0's in front of the first half of these numbers and 1's in front of the last half: 000, 001, 011, 010, 110, 111, 101, 100. This corresponds to a Hamiltonian path starting at 000 on a cube.

Proceeding in this fashion, first reflecting the entire series, then adding 0's and 1's on the left, one can quickly generate the reflected binary Gray code to

	FEDCBA		FEDCBA
0	0⎫	21	1 1 1 1 1⎫
1	1⎭	22	1 1 1 0 1⎭
2	1 1	23	1 1 1 0 0
3	1 0	24	1 0 1 0 0⎫
4	1 1 0	25	1 0 1 0 1⎭
5	1 1 1⎫	26	1 0 1 1 1
6	1 0 1⎭	27	1 0 1 1 0
7	1 0 0	28	1 0 0 1 0
8	1 1 0 0⎫	29	1 0 0 1 1⎫
9	1 1 0 1⎭	30	1 0 0 0 1⎭
10	1 1 1 1	31	1 0 0 0 0
11	1 1 1 0	32	1 1 0 0 0 0⎫
12	1 0 1 0	33	1 1 0 0 0 1⎭
13	1 0 1 1⎫	34	1 1 0 0 1 1
14	1 0 0 1⎭	35	1 1 0 0 1 0
15	1 0 0 0	36	1 1 0 1 1 0
16	1 1 0 0 0⎫	37	1 1 0 1 1 1⎫
17	1 1 0 0 1⎭	38	1 1 0 1 0 1⎭
18	1 1 0 1 1	39	1 1 0 1 0 0
19	1 1 0 1 0	40	1 1 1 1 0 0⎫
20	1 1 1 1 0	41	1 1 1 1 0 1⎭
		42	1 1 1 1 1 1

Figure 4   Reflected binary Gray code for 0 through 42.

any desired counting number. Note that for each set of *n*-tuplets the code is cyclic in that the first and last *n*-tuplets also differ at only one spot. If the code is used by a counter consisting of wheels, such as the usual mileage meter, the meter can go from its highest number back to 0's with a final unit change of only one wheel.

In 1872 Louis Gros published in Lyon a brochure on *Théorie du Baguenodier*. "Baguenodier" (more commonly spelled "baguenaudier") is the French name for a classic puzzle known in the English-speaking world as Chinese rings, although any connection between the puzzle and China is unknown to me. In his brochure Gros applied a binary notation to this puzzle for the first time. The puzzle had been first described in 1550 by Girolamo Cardano in his *De Subtilitate Rerum,* and it was later analyzed at considerable length by John Wallis in his *Algebra* in 1693.

Many versions of the Chinese rings (the number of rings can vary) are currently on sale around the world. If you are handy with tools, the puzzle can be made with curtain rings, stiff wire and a strip of wood with holes drilled through it *[see Figure 5].*

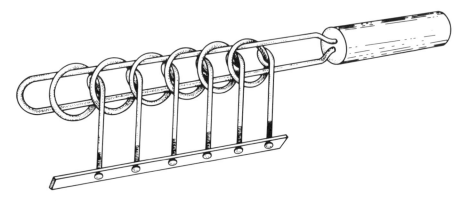

Figure 5   Chinese ring puzzle.

The object of the puzzle is to free all the rings from the double bar. For a first move the two end rings can be dropped either individually or both at once. To simplify the solution, we shall assume that only one of the two end rings is dropped at a time. With the exception of those two rings (which can always be taken off or put on simultaneously), a ring will come off or go on only when its immediate neighbor closer to the end is on and all the other rings beyond are off. This is the peculiar feature of the puzzle that makes it so frustrating and repetitious.

Let each ring be represented by a binary digit: 1 for on, 0 for off. The binary Gray number for 42 [*see Figure 6*] is 111111. If we let this represent the six rings on the upper rod, each Gray number going from 42 back to 0 shows which ring is to be removed or put on to solve the puzzle in a minimum number of moves! For *n* rings it is apparent that to determine the number of moves required, we simply write *n* as a Gray number of *n* units, convert it to standard binary and so obtain the answer. In this case the Gray number 111111 corresponds to 101010 in standard binary, which is 42 in decimal notation. (Gros explained all this in a slightly different way, but it amounts to the same thing.) To find the number by formula, use $\frac{1}{3}(2^{n+1} - 2)$ when *n* is even and $\frac{1}{3}(2^{n+1} - 1)$ when *n* is odd.

We have assumed that for each move only one ring is removed or put on. The braces in Figure 4 indicate pairs of moves that can be made simultaneously with the two end rings. If these are counted as single moves, the six-ring puzzle can be solved in 31 moves instead of 42. The formulas for this "fast way" of solving an *n*-ring puzzle are $2^{n-1} - 1$ if *n* is even and 2^{n-1} if *n* is odd.

With a six-ring puzzle the slow-to-fast ratio is 42/31 = 1.355; for seven rings it is 85/64 = 1.328. The ratios continue as 1.338, 1.332, 1.334,. . . . N. S. Mendelsohn has shown that this oscillating series converges rapidly to $1\frac{1}{3}$.

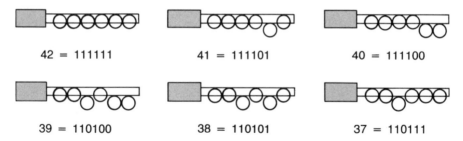

42 = 111111   41 = 111101   40 = 111100

39 = 110100   38 = 110101   37 = 110111

Figure 6   First six positions for solving ring puzzle using the Gray code.

Twenty-five rings require 22, 369, 621 steps. Assuming that a skilled operator can do 50 steps a minute, he could solve the puzzle the slow way, working 10 hours per day, in a little more than two years. By doing it the fast way, however, he could cut the time by about half a year.

Jesse R. Watson of Altadena, Calif., headed a firm called Watson Products that manufactured a handsome, six-ring, aluminum version of the rings in the early 1970's. In his instructions he asked the following question: Suppose the initial position for an n-ring puzzle has the last ring (the one nearest the handle) on and all other rings off. Watson calls this the position of "maximum effort" because it requires more moves than any other position to take all the rings off. Assuming that the slow method is used, what simple formula gives the required minimum number of moves?

The binary Gray code also solves the well-known Tower of Hanoi puzzle, in which n disks of diminishing sizes are stacked in a pyramid. The problem is to transfer them one at a time to a second spot, using a third spot as a temporary resting place with the proviso that no disk be placed on top of a smaller disk. (See Chapter 6 of *The Scientific American Book of Mathematical Puzzles & Diversions.*) To solve this puzzle for five disks, label the disks of the initial pyramid, starting with the smallest, from A to E. Label the columns of Figure 4 from A to F as shown. Take the Gray numbers in sequence. At each step move the disk that corresponds to the column in which there is a change of digit. The sequence begins *ABACABAD*. . . . On every move a disk can be transferred to only one spot. The sequence solves the puzzle in $2^n - 1$ moves, which in this case is 31.

Rules for converting numbers in other base systems to reflected Gray numbers are simple generalizations of the rules for binary numbers. (There are several general conversion procedures, but I give the simplest here.) If the base is even, the rules are the same as for the binary system, except that when a digit is altered it is changed to its "complement" with respect to $n - 1$ when n is the base, that is, to its difference from $n - 1$. In the binary system, $n - 1 = 1$, so

	GRAY		GRAY
0	0	16	13
1	1	17	12
2	2	18	11
3	3	19	10
4	4	20	20
5	5	21	21
6	6	22	22
7	7	23	23
8	8	24	24
9	9	25	25
10	19	26	26
11	18	27	27
12	17	28	28
13	16	29	29
14	15	30	39
15	14		

Figure 7   Reflected decimal Gray code.

that this means a simple change of 0 to 1 or 1 to 0. In the decimal system, numbers are complemented with respect to 9 (that is, subtracted from 9). Therefore to convert a decimal number to a Gray number take each digit in turn beginning at the right. If the next digit to the left is even, leave the former digit unchanged. If the left digit is odd, complement the former digit. For example, 1972 becomes 1027. To convert back to the decimal system, work with sums. If all digits to the left have an even sum, let the digit stand. If the sum is odd, subtract the digit from 9.

Only a slight modification of rules is required for numeral systems with an odd base. In such cases the sum rule applies to conversion in either direction. In the ternary system, for instance, complementation is with respect to 2. Regardless of which way you convert, complement when the sum on the left is odd; otherwise let the digit stand. Ternary Gray numbers, in counting order, are 0, 1, 2, 12, 11, 10, 20, 21, 22, 122, 121, 120, ...

In all bases, Gray counting numbers of the reflecting type (unless otherwise specified, these are considered *the* Gray numbers for a given base) are quickly determined by generalizing the procedure given for binary numbers. This is best explained by using the decimal system as an example *[see Figure 7]*. Note that the unit's column begins with the sequence 0 through 9; then it proceeds from 9 through 0, then from 0 through 9 and so on. In the 10's column, ten 0's (not shown) are followed by ten 1's, then by ten 2's, ten 3's and so on through ten 9's until 99 is reached. Now the doublets are reflected after every 100 steps, and in the third column from the right a hundred 0's are followed by a hundred 1's, then by a hundred 2's and so on until 999 is reached. The reader should have little difficulty applying this procedure to other base systems. In the ternary system, for example, reflections occur in the right column every third step, in the next column every ninth step, in the next column every 27th step and so on through increasing powers of 3.

Because Gray codes are relatively unknown to students of recreational mathematics, I suspect they have many puzzle applications other than the ones given here. I would be glad to hear from readers who know of recreational uses for Gray codes with bases greater than 2.

ANSWER

From a "maximum effort" position (only the last ring is on the bar), $2^n - 1$ moves are required to remove all the rings by the slow method. Numbers of this form are called Mersenne numbers. The same formula gives the number of moves required for transferring n disks in the Tower of Hanoi puzzle.

Henry E. Dudeney, in his discussion of the puzzle (see the bibliography), has this to say about a "maximum position" task. "If there are seven rings and you take off the first six, and then wish to remove the seventh ring, there is no course open to you but to reverse all those 42 moves that never ought to have been made. In other words, you must replace all the seven rings on the loop and start afresh!"

ADDENDUM

The limerick at the head of the chapter is only half anonymous. It is my variation on the following anonymous tribute to the binary system:

> The binary system is fun,
> For with it strange things can be done.
>   Two as you know
>   Is a one and an oh,
> And five is one hundred and one.

Like so many mathematical ideas, the origin of the Gray code fades into history. George R. Stibitz, a physiologist at Dartmouth Medical School, sent me a copy of his 1943 patent (No. 2,307,868), applied for in 1941 when he was with Bell Laboratories. It describes a counting device using elastic balls and magnets. Electric pulses shift the balls back and forth, varying their positions in accord with the cyclic Gray code. Recalling this patent prompted Stibitz to write:

> An ingenious fellow one day
> Wrote numbers a new-fangled way.
>   As earlier had Stibitz,
>   But that name inhibits
> Historians who call the code "Gray."

So far as I have been able to learn, the earliest technical use of the Gray code was by Émile Baudot (1845 – 1903), a French engineer who applied the cyclic code to telegraphy. For details and references see "Origins of the Binary code," by G. G. Heath in *Scientific American,* August, 1972, pages 76 – 83.

The term "reflected code" was first used by Gray in his 1953 patent. "Because this code in its primary form may be built up from the conventional binary code by a sort of reflection process and because other forms may in turn be built up from the primary form in similar fashion, the code in question, which has as yet no recognized name, is designated in this specification and in the claims as a 'reflected binary code.' "

Sydney N. Afriat, an economist at the University of Ottawa, has written an entire book about the Chinese rings, *The Ring of Linked Rings,* published in London in 1982 by Duckworth and Company. Afriat also discusses the Tower of Hanoi and how the two puzzles are solved by the Gray code. The book includes computer programs for both puzzles and an extensive bibliography.

In recent years several mechanical puzzles have been marketed that use a Gray code for their solution. A notable example is The Brain, invented by computer scientist Marvin H. Allison, Jr., and made in the 1970's by a company called Mag-Nif. It consists of a tower of eight transparent plastic disks that rotate horizontally around their centers. The disks are slotted, with eight upright rods going through the slots. The rods can be moved to two positions, in or out, and the task is to rotate the disks to positions that permit all the rods to be moved out. The Gray code supplies a solution in 170 moves.

A curious puzzle called Loony Loop — its complicated history would require a chapter — consists of four intertwined steel loops and a ring of nylon cord that seems permanently captured by the loops. The task is to free the nylon cord. The puzzle generalizes to n metal loops and is solved by applying a ternary Gray code to a sequence of moves.

Many readers reminded me of the similarity of Gray codes to a word puzzle invented by Lewis Carroll, called Doublets, better known today as Word Ladders. The idea is to change a word to one of the same length by altering one letter at a time, forming a different word at each step, in a minimum number of steps. (See the chapter on Carroll in my *New Mathematical Diversions from Scientific American.*) As often noted, word ladders resemble the way the genetic code is altered by evolutionary mutations. On the relation of the Gray code to word-ladder problems, see "The Arithmetic of Word Ladders," by Rudolph W. Castown, in the quarterly journal *Word Ways,* Vol. 1, August, 1968, pages 165 – 169.

The Gray code solves many brainteasers that appear from time to time. A typical example is the switching puzzle on page 26 of *The Surprise Attack in Mathematical Problems,* by L. A. Graham (Dover, 1968). An earlier instance is Problem 319, solved in *American Mathematical Monthly,* December, 1938, pages 694 – 696. Imagine a light bulb connected to n switches in such a way that it lights only when all the switches are closed. A push button opens and closes each switch, but you have no way of knowing which push opens and which closes. What is the smallest number of pushes required to be certain you will turn on the light regardless of how the switches are set at the outset? This device, by the way, is the basis of an amusing trick, unpatented and inventor unknown, that is currently sold in magic stores under various trade names. Louis Tannen's Magic Studio in Manhattan sells it under the name Electronic Monte. There are three push buttons and one light. The magician demonstrates how a single button seems to control the light, but the control mysteriously changes from one push button to another, like the pea in a three-shell game.

Many legends tell how the rings puzzle was invented in ancient China, but the world's expert on early Chinese inventions, Joseph Needham, finds no evidence of its Asian origin. (See his *Science and Civilization in China,* Vol. 3, page 111.) The Japanese became so intrigued by the puzzle in the 17th century that they wrote Haiku poems about it, and symbols of the linked rings appeared on heraldic emblems. There is a large literature on the puzzle in both China and Japan, but I know of no published bibliography.

The rings were sometimes used in Europe as a whimsical locking device for bags and chests. In England the puzzle was usually called the "tiring irons," probably because it is tiring to solve, especially if the rings are large and heavy. According to the Oxford English Dictionary, it was earlier called "tarrying irons," perhaps because one is long delayed in solving it. Among its quotations, the OED gives the following 1782 doggerel:

> Have you not known a small machine
> Which brazen rings environ,
> In many a country chimney seen
> Y-clept a tarring-iron?

The puzzle has been sold around the world in hundreds of forms. A beautiful hand-carved ivory version with nine rings (from the puzzle collection of Tom Ransom of Toronto) provided the cover for the May 1977 issue of *Computer*. The puzzle illustrates the last-in – first-out principle of stack machines, the topic of five articles in the issue. At the other extreme is a small seven-ring version I found advertised in a 1936 Johnson Smith and Company catalog, where it is called the "Chinese Ringbar Puzzle"; price: 15 cents. "This is an extremely difficult puzzle," the description reads, "yet very simple when you are familiar with the method. . . . You may try forever and not be able to remove the bar from the rings. Just as you think you are getting it done, you are further off the solution than ever, and you have to give up in despair."

An elaborate electronic version of the puzzle, with eight lights and eight push buttons, is the topic of "The Princeps Puzzle," by James W. Cuccia, in *Popular Electronics*. The article gives detailed instructions on how to make the thing. (I am indebted to Dr. Burton J. Bacher for calling this article to my attention.)

Science and Invention, September, 1927, page 397, describes a "marvelous escape trick" in which a "fair damsel" is shackled on the stage in the manner shown in Figure 8. Diagrams show how the lady is released by manipulating the rings around her arms and legs in the manner of the Chinese puzzle. This preposterous stage trick, the article says, was invented by one Theodore P. Brunner of Los Angeles, who has it protected by U.S. Patent 1,625,452.

Since my column on Gray codes was published in *Scientific American* in 1972, the number of 5-bit codes — the same as the number of Hamiltonian paths on a five-dimensional cube — has been determined. A good upper bound has been established for the 6-bit Gray code.

At the 1980 IEEE International Conference on Circuits and Computers, at Port Chester, N.Y., a paper was presented titled "Gray Codes: Improved Upper Bounds and Statistical Estimates for $n > 4$ bits," and published in 1983 (see the bibliography). The authors were Jerry Silverman, Virgil E. Vickers and John L. Sampson, electrical engineers at the Rome Air Development Center, Hanscom Air Force Base, Chicopee, Mass. Their estimates for the 5- and 6-bit codes were based on Monte Carlo techniques.

The authors open with a succinct definition of an n-bit Gray code as "a list of all the 2^n binary n-tuples ordered so that adjacent elements differ by a change in only one bit." They point out that such codes are widely used in A/D conversion, shaft encoding, codes for data retrieval, control mechanisms, switching and network theory and experimental design. Although the reflected binary code is the most widely used, other types of Gray codes are preferred for special purposes. Finding a formula for the number of n-bit Gray codes as a function of n remains a difficult unsolved combinatorial problem.

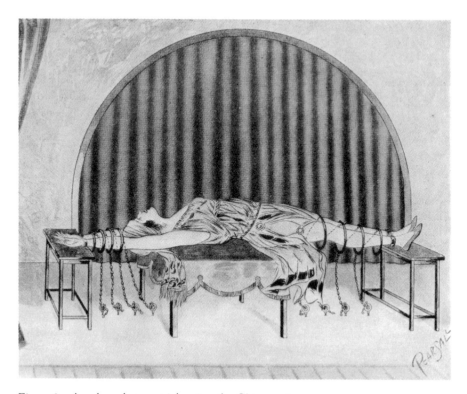

Figure 8   An absurd stage trick using the Chinese rings.

Their statistical estimates agreed with the exact value for the 4-bit code to within 0.06 percent. To test their estimate for the 5-bit code they made a precise calculation on a PDP-11 computer. At first they feared the running time would be about 11 years, but by taking advantage of symmetries and a "look ahead" method that predicts dead-end branches, they were able to reduce the running time to 750 hours. "We are happy to report," the three researchers told me in a 1980 letter, "that a five-dimensional fly can walk along the edges of a five-dimensional cube in exactly 187,499,658,240 ways."

We need to make clear just what this number counts. It allows the fly to start at any corner of the hypercube and trace a Hamiltonian path that ends at any other corner. Reversals of each path are included. (If reversals are excluded, the number must be halved.) The number of Hamiltonian circuits — paths starting anywhere but ending on a corner adjacent to the starting corner — is 58,018,928,640. This, too, includes reversals. If you want the corresponding figures for paths and circuits that start only at the corner taken as zero, then each of the above figures must be divided by $2^n = 32$, where n is the dimension number.

Because my column did not disclose the numbers for Hamiltonian paths and circuits (including reversals) on lower-order cubes, I give them below:

n	Hamiltonian circuits	Hamiltonian (noncyclic) paths
1	2	0
2	8	0
3	96	48
4	43,008	48,384

The Hanscom researchers were the first to publish the figures for the 5-bit Gray code, but they were not the first to find them. After my column appeared, the same results were sent to me in the fall of 1972 by David Vanderschel of Houston, Alex G. Bell and Peter Hallowell of the Rutherford High Energy Laboratory in Chilton, England, and Steve Winker of Naperville, Ill. None published their results, but I reported them in my column for April, 1973. The fact that all four programs agreed is strong evidence that the numbers are accurate.

The number of 6-bit Gray codes remains unknown. The Hanscom researchers estimate it as close to 2.4×10^{25}, a number so large that it is probably not possible to determine it precisely in any reasonable computer time, unless, of course, someone discovers a formula or some new algorithm shortcuts.

I don't know if anyone has noticed this before, but it occurred to me that the number of Gray ternary codes of n digits is equal to the number of Hamiltonian paths on n-cubical lattices with three points on each edge and the faces toroidally joined. This is best explained with examples.

There are six one-digit ternary Gray codes. We represent them on the single edge of a one-dimensional "cube" by three points, then close the ends of the line to make a circle, as shown in Figure 9a. By starting at any point and counting reversals, we see that the six Hamiltonian paths provide the one-digit codes 012, 120, 201 and their reversals; 210, 021 and 102.

The two-digit ternary Gray codes are obtained from the square lattice shown in Figure 9b. Its nine points are labeled with the nine two-digit combinations of 0, 1 and 2. Points on each side of the square are connected to the three points on the opposite side. The graph can, of course, be drawn on a torus with three parallel lines going around it one way and three circling it the other way. The number of Gray codes is the number of Hamiltonian paths on this graph. Already it is not easy to see how to count the paths systematically, and I have made no attempt to do so.

For three-digit ternary codes we go to the cubical lattice shown in Figure 9c, its 27 points labeled with the 27 three-digit combinations of 0, 1 and 2. As

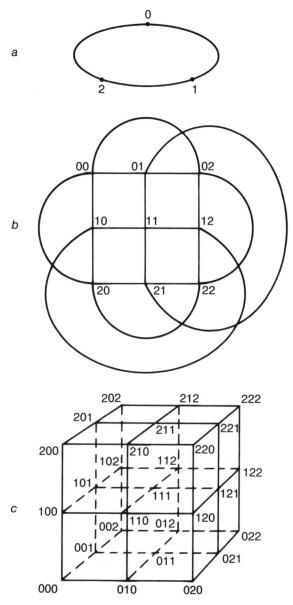

Figure 9   Ternary Gray codes as Hamiltonian paths.

before, imagine each point connected by a line (not shown) to the corresponding point on the opposite face. The procedure clearly generalizes to n-dimen-
sional hypertoruses. Gray codes with bases higher than 3 can be similarly
generated by Hamiltonian paths on more complicated hyperlattices.

BIBLIOGRAPHY

Gray Codes

"Reflected Number Systems." Ivan Flores in *IRE Transactions on Electronic Computers*, Vol. EC-5, 1956, pages 79–82.

"Affine *M*-ary Gray Codes." Martin Cohn in *Information and Control*, Vol. 6, 1963, pages 70–78.

"Digital Transmission of Analog Signals." William R. Bennett in *Introduction to Signal Transmission*. McGraw-Hill, 1970.

"Using the Decimal Gray Code." N. Darwood in *Electronic Engineering*, February, 1972, pages 28–29.

"On the Use of Binary and Gray Code Schemes for Continuous-Tone Picture Transmission." E. S. Deutsch in *Pattern Recognition*, Vol. 5, 1973, pages 121–132.

"Distance-2 Cycle Chaining of Constant Weight Codes." D. T. Tang and C. N. Liu in *IEEE Transactions on Computers*, Vol. 22, 1973, pages 176–180.

"Efficient Generation of the Binary Reflected Gray Code and Its Applications." J. R. Bitner, G. Ehrlich and E. M. Reingold in *Communications ACM*, Vol. 19, 1976, pages 517–521.

"Gray Codes." Edward M. Reingold, Jurg Nievergelt and Narsingh Deo in *Combinatorial Algorithms*, pages 173–188. Prentice-Hall, 1977.

"A Technique for Generating Gray Codes." J. E. Ludman and J. L. Sampson in *Journal of Statistical Planning and Inference*, Vol. 5, 1981, pages 171–180.

The Chinese Rings

"*Le Jeu du Baguenaudier.*" Edouard Lucas in *Récréations Mathématiques*, Vol. 1, Chapter 7. Paris, 1883.

"*Der Baguenaudier.*" W. Ahrens in *Mathematische Unterhaltungen und Spiele*, Vol. 1. Berlin: Druck und Verlag von B. G. Teubner, 1910.

"The Tiring Irons." H. E. Dudeney in *Amusements in Mathematics*, Problem 417. Nelson, 1917.

"Some Binary Games." R. S. Scorer, P. M. Grundy and C. A. B. Smith in *Mathematical Gazette*, Vol. 28, 1944, pages 96–103. The rings puzzle is generalized to a tier of *k* rods.

"Chinese Rings." Maurice Kraitchik in *Mathematical Recreations*. Dover, 1953.

"The Icosian Game and the Tower of Hanoi." Martin Gardner in *The Scientific American Book of Mathematical Puzzles and Diversions*. Simon and Schuster, 1959.

"Problems and Puzzles." Joseph Needham, *Science and Civilization in China*, Vol. 3, Section 19. Cambridge University Press, 1959.

"An Old Puzzle." E. H. Lockwood in *Mathematical Gazette*, Vol. 53, 1969, pages 362 – 364.

"The Princeps Puzzle." James W. Cuccia in *Popular Electronics*, Vol. 34, 1971, pages 26 – 32.

"Chinese Rings." W. W. Rouse Ball in *Mathematical Recreations and Essays*, 12th edition, edited by H. S. M. Coxeter. University of Toronto Press, 1974.

Hamiltonian Paths on the n-Cube

"Gray Codes and Paths on the n-Cube." E. N. Gilbert in *The Bell System Technical Journal*, Vol. 37, 1958, pages 815 – 826.

"Graph Theory Algorithms." Ronald C. Read in *Graph Theory and Its Applications*, edited by Bernard Harris. Academic, 1970.

"Cyclic Codes in Analog-to-Digital Encoders." Charles F. Cole, Jr., in *Computer Design*, May, 1971, pages 107 – 112. Shows how Gray codes can be counted by tracing Hamiltonian paths on two-dimensional matrices known in network theory as Karnaugh maps.

"A Technique for Generating specialized Gray Codes." Virgil E. Vickers and John L. Silverman in *IEEE Transactions on Computers*, Vol. C-29, 1980, pages 329 – 331.

"Statistical Estimates of the n-Bit Gray Codes by Restricted Random Generations of Permutations of 1 to 2^n." Jerry Silverman, Virgil E. Vickers and John L. Sampson in *IEEE Transactions on Information Theory*, Vol. IT-29, 1983, pages 894 – 901.

"A Cube-Filling Hilbert Curve." William J. Gilbert in *Mathematical Intelligencer*, Vol. 6, 1984, page 78. Shows how the 3-bit reflective binary Gray code can be used to start a sequence that generates a Hilbert curve that will at the limit completely fill an n-dimensional cube.

CHAPTER THREE

Polycubes

In 1958 Piet Hein's Soma cube was first introduced to U.S. puzzle buffs in my September *Scientific American* column. (The column is reprinted in *The 2nd Scientific American Book of Mathematical Puzzles & Diversions*). The puzzle has since been sold around the world under a variety of trade names. The only authorized version is marketed in the U.S. by Parker Brothers, with an informative booklet written and illustrated by Piet Hein. Three issues of *Soma Addict,* a newsletter edited by Thomas V. Atwater, were published, as well as many articles on Soma in mathematical journals.

The Soma pieces are a subset of what have been called polycubes. These are solid figures created by joining unit cubes at their faces. Like their flat cousins the polyominoes, they pose an extraordinarily difficult combinatorial problem: Given *n* cubes, is there a formula for calculating the number of distinct polycubes of order *n*? If so, it has not yet been found, although there are, of course, recursive procedures by which all polycubes of order *n* can be constructed: Simply add a cube in all possible ways to each polycube of order *n* − 1 and eliminate duplicates. Since there is no way to "turn over" an asymmetric polycube in 4-space analogous to the way an asymmetric polyomino can be reversed in 3-space, mirror-image pairs of polycubes are considered different. It is obvious that for orders 1 and 2 only one polycube is possible for each and that three unit cubes can form two polycubes. It also is easy to determine that there are eight tetracubes and 29 pentacubes. Several computer programs have verified a hand computation, first made by David Klarner, that there are 166 hexacubes. As far as I am aware, the number of heptacubes is still undetermined.

The Soma cube consists of the seven irregular shapes *[See Figure 10]* that can be formed by combining three or four unit cubes — all nonconvex polycubes of orders 1 through 4. There are 240 distinct ways (not counting rotations and reflections) that the seven pieces will form a 3-by-3-by-3 cube. This was first

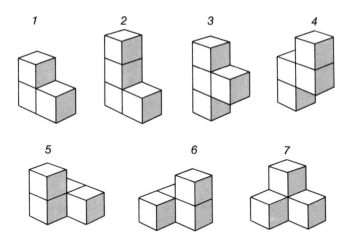

Figure 10   The seven Soma pieces.

determined by John Horton Conway and M. J. T. Guy and has since been verified by many computer programs. Parker's Soma booklet states that Conway and Guy used a computer for their work — an error I am now happy to correct. As Conway puts it in a letter, he and Guy, both mathematicians at the University of Cambridge, obtained the 240 solutions by hand "one wet afternoon" when they had no more-pressing chores.

"I think for a puzzle the size of Soma," Conway adds, "it's an admission of defeat to use a computer. If you find the right way of organizing the material, it should take less time to do the whole thing by hand than it does to program the machine." By first establishing a few ingenious theorems (some of which were found by Guy's father, R. K. Guy) and using a parity coloring technique, they were able to check all possibilities with great efficiency.

Conway and Guy later discovered that if you begin with any of 239 solutions (one solution is an anomaly), all of the others can be obtained in 238 steps by altering the position of no more than three pieces at each step. Conway has drawn a large graph (which he calls the Somap) showing how the 239 solutions are linked to one another and giving each solution a concise notation, called its "somatype." The map does not give any one solution, but once you have built the cube in any of the 239 ways, the map enables you to transform it to all the others by moving two or three pieces at a time. The map is too complex to reproduce here, but you will find it on pages 802–803 of *Winning Ways*, Vol. 2, by Elwyn R. Berlekamp, John H. Conway and Richard Guy (Academic, 1982).

The Soma cube's popularity flows from the enormous variety of pleasing shapes that can be made with its pieces and from the many clever ways of proving that certain 27-cube shapes are impossible. It is not, however, the first

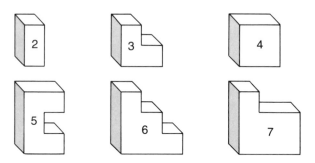

Figure 11   Polycube pieces for the Diabolical cube.

polycube dissection of the order-3 cube to be marketed as a puzzle. A six-piece set was sold in Victorian England under the name of the Diabolical cube [see Figure 11, top]. (Its pieces are reproduced on page 108 of *Puzzles Old and New*, by "Professor Hoffmann," published in London in 1893.) I do not know how many basic solutions the Diabolical cube has, but perhaps a reader can tell me. I found only eight. The pieces can be cut from wood or made by gluing together alphabet blocks. As Piet Hein has noted, the unknown inventor surely intended a dissection of the cube into a set of "flat" polycubes containing one each of orders 2 through 7.

Another dissection of the cube into six polycubes was made by the Polish mathematician J. G. Mikusiński [see Figure 12, middle]. It appears in Hugo Steinhaus's *Mathematical Snapshots* (Oxford University Press, 1950). These pieces are currently on sale here and abroad under several trade names. There are just two solutions, both difficult to find. Still another interesting cube dissection, suggested by Thomas H. O'Beirne of Glasgow, is to cut the order-3 cube into nine tricubes, all shaped like the 3-piece of the Diabolical cube. Random attempts to build a cube with the nine tricubes are likely to be very frustrating unless you hit on a systematic procedure.

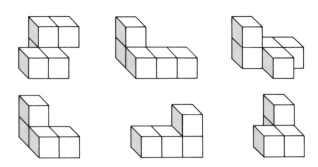

Figure 12   Polycube pieces for J. G. Mikusinski's cube.

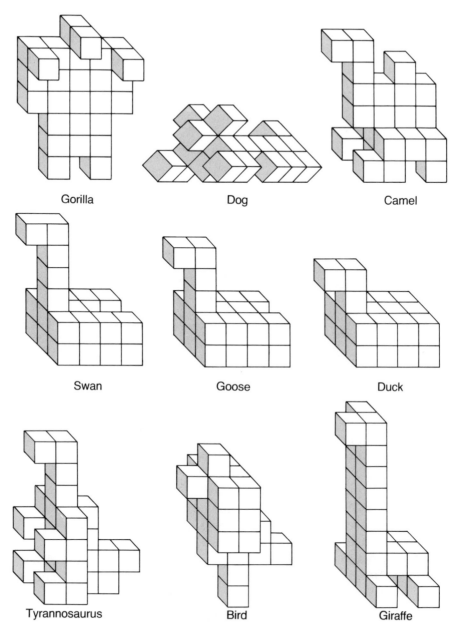

Figure 13 Soma animals created by Rev. John W. M. Morgan.

Nine animals from a zoo of several dozen Soma figures created by Rev. John W. M. Morgan, vicar of St. Matthew's Church in Luton, England, are shown in *Figure 13*. The animals all have bilateral symmetry except for the giraffe, whose head leans to one side (he is thinking), and the dog, whose hidden rear portion violates symmetry. The bird actually will perch on one leg as shown.

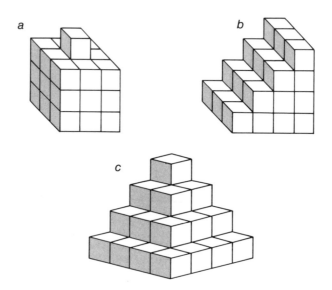

Figure 14   Soma structures with hidden holes: (*a*) penthouse,
(*b*) staircase and (*c*) tower.

Three Soma structures of a delightful new type were created by Benjamin L. Schwartz of McLean, Va. *[see Figure 14, bottom]*. The penthouse has a cubical hole at its center and is not hard to construct. The tower is flat on its two hidden sides and has three interior holes. The stairway also has three interior holes. The last two are difficult to build. In both cases the three holes are inside, invisible from all angles.

Another pleasant exercise is to construct Schwartz's three figures with the six Diabolical-cube pieces. None is possible with the holes on the interior, but each can be made with one or more holes at the back, so that the structures appear as shown in the illustration.

The notorious wall in Piet Hein's instruction booklet is an insoluble Soma problem *[see Figure 15]*. Many impossibility proofs have now been found, but

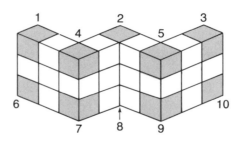

Figure 15   The impossible wall.

the simplest (discovered independently by many Soma addicts) is based on the wall's 10 corner cubes, shown shaded in the illustration. If each Soma piece is considered in turn, it is apparent that five of the pieces can provide only one corner cell each and that the other two can provide no more than two each. All together, therefore, the pieces can supply a maximum of nine corners. Since there are 10 corners, the wall is impossible. It *is* possible, however, to build a wall that from the front looks exactly like the one in the illustration. If the wall is viewed from behind, however, the hidden corner (indicated by the arrow) is missing, and an extra cube protrudes at some other spot.

The corner proof of impossibility applies also to the six pieces of Mikusiński's cube but not to the Diabolical pieces. Unfortunately, they will not make a genuine wall either, and readers may enjoy proving it by a different technique. The Diabolical pieces will, however, like the Soma pieces (but not Mikusiński's), make an ersatz wall that appears genuine from the front. This is a harder task than forming the Diabolical cube. There are several ways to do it with one hole hidden below the top center corner and one back-projecting cube at the base, where it is hard to see even when looking downward from the front. A not-so-funny joke to play on a victim is to let him see a false wall from the front (formed by either Soma or Diabolical pieces), knock the wall apart and then offer him $50 if he can rebuild the structure (with no holes, of course) within three hours.

The building of fake structures opens up numerous amusing possibilities. One can build Soma bricks that are 3 by 3 by 4 or 2 by 3 by 6, that appear solid but are actually hollow in back like the façades of buildings on a movie set. A spurious 2-by-2-by-8 tower can even display two extra cubes on top. A 1-by-4-by-6 Soma wall, standing on edge, has three invisible cubes projecting from the back. Of course, gravity must be taken into account in problems of this type because the structures should be capable of standing alone, without the aid of adhesive or concealed supports.

Many people have worked on structures formed from larger sets of polycubes. The eight tetracubes were manufactured in Hong Kong in 1967 (by E. S. Lowe Co., Inc.) and marketed as the Wit's End puzzle. The set came boxed as a 2-by-2-by-8 solid. A 2-by-4-by-4 solid also is possible. Indeed, a group in the Artificial Intelligence Laboratory at the Massachusetts Institute of Technology used a computer to show that it had 1,390 basic solutions. Both of these solids are enlarged replicas of two of the tetracubes, and enlarged replicas of the remaining six tetracubes can also be made.

The 29 pentacubes are the subject of U.S. Patent 3,065,970, November 27, 1962, issued to Serena Sutton Besley. Unfortunately, no rectangular solid has $5 \times 29 = 145$ unit cubes, but by adding a duplicate pentacube, Mrs. Besley

Figure 16　The solid pentominoes.

obtained 150 unit cubes. The 30 pieces will form bricks of 5 by 5 by 6, 3 by 5 by 10, 2 by 5 by 15 and 2 by 3 by 25. Klarner had earlier found that if the 1-by-1-by-5 piece is omitted, the remaining 28 pentacubes will form two separate 2-by-5-by-7 solids. Two solutions are given by Solomon W. Golomb in his *Polyominoes,* page 118. Other problems devised by Klarner, using 28 or fewer pentacubes, are in Golomb's book on pages 159–160.

If the 12 pentominoes are given a unit thickness, the set is known as the solid pentominoes [*see Figure 16*]. Golomb introduces this popular set of polycubes on page 116 of his book and gives additional problems with them on pages 158–159. The set will form enlarged replicas of nine of the pieces. When Golomb's book appeared, the W and X pieces had been proved impossible, but replicating the F piece (sometimes called the R piece) remained undecided until 1970. It was solved by J. M. M. Verbakel of the Philips Research Laboratories in the Netherlands. It is not known if his solution (see the bibliography) is unique.

C. J. Bouwkamp, associated with the same laboratory, reported in 1969 (see the bibliography) on his computer programs that produced all the solutions for packing the 12 solid pentominoes in boxes of 2 by 3 by 10, 2 by 5 by 6 and 3 by 4 by 5. The number of basic solutions is 12, 264 and 3,940, respectively. Bouwkamp's paper gives the 12 solutions for the 2 by 3 by 10 and comments on some of their unusual properties. In July, 1967, the Technological University of Eindhoven published Bouwkamp's 310-page *Catalogue of Solutions of the Rectangular 3 × 4 × 5 Solid Pentomino Problem.*

An unusual task that links the solid pentominoes with the Soma puzzle has been proposed by J. Edward Hanrahan of La Mesa, Calif. He reports that it is possible to form 4-by-4-by-2 solids with the Soma pieces so that on the upper 4-by-4 layer there will be five cubical holes joined to form hollow molds for each of the 12 solid pentominoes except the I pentomino, which is obviously too long to fit.

Working with cubical holes suggests many curious and unsolved polycube questions. What, for example, is the largest volume of empty space that can be put inside a solid formed with a specified set of polycubes? "Inside" can be defined in various ways. What is the maximum number of unit holes that do not touch one another or touch the outside surface (under various definitions of "touch")?

Here is an intriguing, unpublished and unsolved hole problem that can be worked on with a set of either the flat pentominoes or the solid ones. Stephen Barr of New York (not Stephen Barr the writer, who lives in Woodstock, N.Y.) recently set himself the task of creating a flat-pentomino pattern having the maximum number of unit holes that do not in any way touch the perimeter or one another. (Each hole must be surrounded by eight squares.) His best result, 12 holes, is shown in Figure 17 in one of several solutions. It can be proved that 14 holes are impossible. I leave it to readers to settle the question of whether or not a pattern with 13 holes can be achieved.

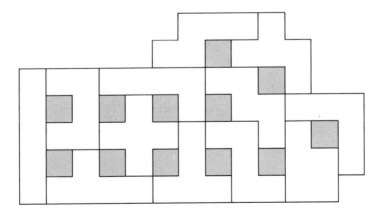

Figure 17 The maximum-hole problem.

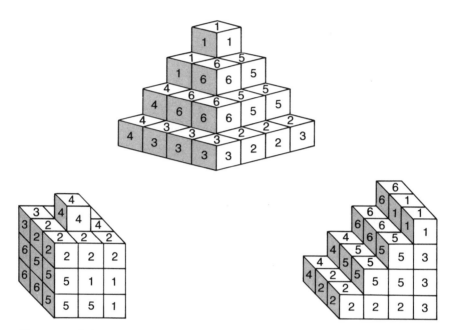

Figure 18   Solutions to the Soma problems.

ANSWERS

Solutions to the Soma tasks of building the penthouse (with one interior hole) and the tower and stairsteps (each with three inside holes) are shown in Figure 18. Numerals indicate the pieces as they are numbered in Figure 10.

Thomas H. O'Beirne's simple procedure for building the 3-by-3-by-3 cube with nine bent tricubes is to use six of them to make three 1-by-2-by-3 slabs. The remaining three tricubes are piled into a stack of height 3; then the slabs are placed vertically *[see Figure 19]*. The picture is a view of the cube from above.

I said that the number of heptacubes had not been calculated. I have since learned from David Klarner and C. J. Bouwkamp that an ALGOL-60 program, written in 1969 by A. J. Dekkers at the Philips Research Laboratories in the Netherlands, found $2^{10} - 1 = 1,023$ heptacubes. This was confirmed in 1972 with a program written by Timothy L. Bock, of Oberlin, Ohio. The results of an earlier program were proved faulty by Klarner's father, who had built a set of wooden heptacubes that included several the program had missed. Klarner assures me that the complete set will pack a 2-by-6-by-83 box, but whether it packs a 3-by-4-by-83 box is not yet known.

Bouwkamp, who also works at the Philips Laboratories, informs me that he wrote a program in 1970, proving that J. M. M. Verbakel's way of replicating

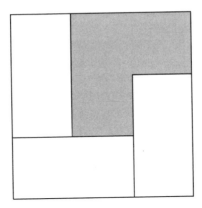

Figure 19   Solution to the tricube problem.

the F-pentacube with the 12 solid pentominoes is unique. "It is understandable," Bouwkamp comments, "that in Golomb's book the replication of this pentacube was left undecided, and most remarkable that Verbakel hit on it by trial and error."

ADDENDUM

Wade E. Philpott, of Lima, Ohio, was the only reader who sent all 13 solutions to the Diabolical cube. I once had occasion to show this puzzle to John Horton Conway of the University of Cambridge. He mentally labeled the pieces with a checkerboard coloring; he then began testing the pieces rapidly, talking out loud and occasionally scribbling a note. It was like watching Bobby Fischer play blitzkrieg chess. About 15 minutes later he announced that there were just 13 solutions. To distinguish them, designate each piece of the Diabolical cube by the number of unit cubes it contains. There are three ways in which the two largest pieces, 6 and 7, can go:

1. Parallel and side by side. When properly placed, with the 5-piece wrapped around a projecting cube of 6, the 4-piece can go in three places. There are five solutions.

2. Parallel but on opposite sides of the cube. There are two solutions.

3. Perpendicular to each other. Crossing in one way yields four solutions, another way two, or six solutions in all.

Philpott also sent a proof that a pattern of 14 unit holes, each surrounded by eight cells, cannot be achieved with the 12 pentominoes. The proof establishes that at least 59 squares are needed to surround 14 holes. On all such patterns

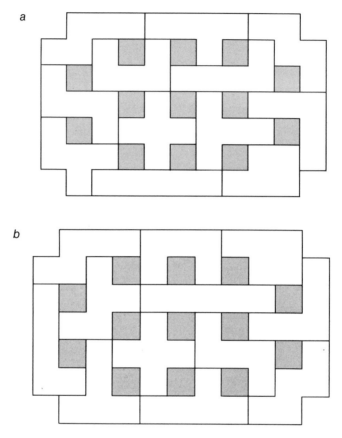

Figure 20   Symmetrical solutions to the 13-hole problem.

each pentomino, except the P and W pieces, will fit. Adding a 60th cell will accommodate only one of the two pieces, proving that the 60 cells of the pentomino set are not enough. Essentially the same proof had earlier been formulated by Joseph Madachy.

Readers too numerous to mention sent 13-hole solutions to the problem. The beautifully symmetrical one shown in Figure 14a was found only by Andrew L. Clarke of Freshfield, England. It so intrigued C. J. Bouwkamp that he wrote a computer program to see if the pattern was unique. He found just one other solution [see Figure 20b], except for rotations and reflections.

If the conditions allow holes to touch the border and also one another at their corners, how many holes are possible? The maximum is 18. The pattern shown in Figure 21, first discovered by Christer Lindstedt of Sweden, may be unique except for a trivial shift of the straight pentomino. If the holes are not restricted

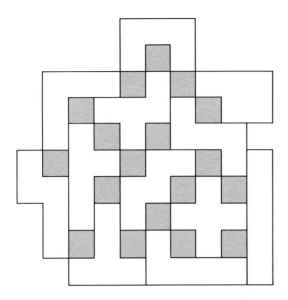

Figure 21 18-unit holes.

to unit squares, there are many 18-hole solutions. (See "Pentomino Problem," in *Journal of Recreational Mathematics*, Vol. 17, No. 3, 1984–1985, pages 220–224.)

The solution I gave for the Soma penthouse with the interior hole is not very stable. John Conway informed me that pieces 4, 5, 6 and 7 can each be used to make the projecting "penthouse," and that the most stable configuration is obtained by forming the cube shown in Figure 22, removing the 7-piece, inverting it, and replacing it. The structure is so stable that if you turn it upside down and put a book on top, it will balance on its projecting cube. This solution was also sent to me by Geof D. Clayton, of Beaverton, Ore.

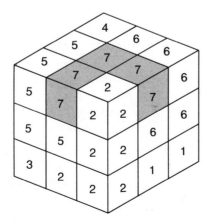

Figure 22 How to make a stable penthouse.

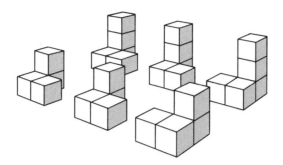

Figure 23   The Lesk cube.

I was wrong in saying that the dog in the Reverend John Morgan's Soma zoo had an asymmetrical rear portion. Peter Neuret of West Germany, sent a symmetrical solution. "My dog was infuriated to read that his hidden rear portion violates symmetry," Morgan wrote. "Just come over here and say that again to his face." Morgan sent two symmetrical solutions.

David Bird of England, raised the interesting question, What is the lowest-order polycube that contains a unit hole completely surrounded? The answer is an order-11 polycube. Six unit cubes are needed to cover the hole's six sides, and five more are required to join them. Note that even if we exclude higher-order polycubes with no interior holes, there are surely structures interlocked in such a way that they cannot be built without going through a fourth dimension. I have no idea what the simplest example would be.

Mathematical Digest, a school periodical in Christchurch, New Zealand, in issue No. 58 (1978) introduced the six polycubes shown in Figure 23. The editors call it the Lesk cube after its designer Lesk Kokay, who had been seeking a dissection of the 3-by-3-by-3 cube into six polycubes that would form the cube in only one way. Unfortunately, it has at least three solutions.

Is there a seven-piece dissection with a unique solution? If so, it has not come to my attention. In 1973 an order-3 cube with, as I recall, seven pieces was on sale in the U.S. under the trade name Qube. The box stated that it had only one solution, but this was achieved by a black-and-white checkerboard coloring of the pieces and the requirement that the cube be similarly colored.

Tom Marlow wrote from England to say that the number of hexacubes and heptacubes were known as far back as 1948. In *The Fairy Chess Review,* Vol. 7, 1948, page 8, Dr. J. Niemann gave the number of heptacubes as 1,023, along with a neat system for classifying them. His figure for the hexacubes was 167, but this was corrected to 166 in a later issue.

Sets of the 12 solid pentominoes have been marketed both here and abroad under various trade names. In the U.S. a handsome polished hardwood set is available from Kadon Enterprises, 1227 Lorene Drive, Pasadena, Md. 21122. It is called Quintillions and comes with a 9-by-12 checkerboard on which games can be played with the pieces. The other 17 pentacubes (those that are not "flat") are also available from Kadon as Super Quintillions. The company also sells "Quint-Art" sculptures, produced by bonding together Quintillion pieces. A four-page *Quint-Gram,* issued twice a year since 1981, is devoted to puzzles based on the solid pentominoes. See the bibliography for other references to pentomino problems.

Joseph Dorrie of Madison Heights, Mich., proposed another subset of the pentacubes — those pieces that are no longer than three units wide along any of the three coordinate directions. There are 25 such pieces, and they form what Dorrie calls the "Dorian cube," a term he has copyrighted.

Another set of polycubes suggested for puzzle purposes consists of all the polycubes in orders 1 through 5. Scott L. Forseth, in "Solid Polyomino Constructions," in *Mathematics Magazine,* Vol. 19, 1976, pages 137 – 139, shows how these 41 polycubes will pack a 2-by-3-by-31 = 186 box. He found two solutions and thinks there are many others.

In 1979 a Los Altos, Calif., firm called Lemmel Associates introduced a puzzle game called Putzl. Invented by L. E. Minnick, it uses two sets of the eight tetracubes, each a different color. One player uses one set; his opponent, the other. They take turns placing one of their pieces on the table. The object is to build an order-4 cube. When a visible face of this cube is completed, the face is won by the player who has the most squares of his color on the face. (The game can be played in reverse, the win going to the person with the least of his color on the face.) A played piece must fit snugly on the previously placed pieces without creating any holes or extending beyond the imagined order-4 cube. If no such play is possible, the player passes. The game ends when no play can be made, and the winner is the person who has captured the most faces. The cube's top face is rarely completed, although the 16 pieces *will* form the cube. Minnick has prepared a handbook for the game.

Lakeside Industries, a division of Leisure Dynamics of Minneapolis, marketed in 1969 a series of six polycube puzzles under the name Impuzzables. Each consisted of a set of five, six or seven plastic polycubes, each set a different color, that fitted together to make a 3-by-3-by-3 cube. The colors were assigned in order of difficulty from the easiest (yellow) to the hardest (blue). Gerard D'Arcey, a California game inventor, designed the puzzles.

Without knowing any of the polycube shapes, or even how many there are to each impuzzable, how quickly can you prove that all the polycubes from all six sets will build a 3-by-6-by-9 brick?

BIBLIOGRAPHY

Polycubes in General

"Solid Polyominoes." S. W. Golomb in *Polyominoes*. Scribner, 1965.

"Packing Boxes with Congruent Figures." D. A. Klarner and F. Göbel in *Koninklijke, Nederlandse Akademie van Wetenschappen. Proceedings, Series A*, Vol. 72, 1969, pages 465–472.

"Symmetry of Cubical and General Polyominoes." W. F. Lunnon in *Graph Theory and Computing*, edited by Ronald C. Read. Academic, 1972.

"Tiling Space with the Aid of the Holomorph." James P. Conlan in *Journal of Combinatorial Theory*, Vol. 14, 1973, pages 167–172.

"Packing Boxes with Congruent Polycubes." Andrew L. Clarke in *Journal of Recreational Mathematics*, Vol. 10, 1977–1978, pages 177–182.

Polycubes. J. Meeus and P. J. Torbijn. Paris: CEDIC, 1917. A marvelous survey of the topic in a book of 176 pages.

Tetracubes

"Tetracubes." Jean Meeus in *Journal of Recreational Mathematics*, Vol. 6, 1973, pages 257–265.

Pentacubes

"Constructions with Pentacubes." N. R. Wagner in *Journal of Recreational Mathematics*, Vol. 5, 1972, pages 266–268.

"Constructions with Pentacubes — 2." *Journal of Recreational Mathematics*, Vol. 6, 1973, pages 211–214.

Pentacubes. Sivy Farhi, published by the author in 1977. The fifth edition (1981) is obtainable from Pentacube Puzzles, Ltd., Box 308, Auckland 1, New Zealand. This is a 70-page booklet of pentacube problems to accompany a set of the 29 pieces. Mr. Farhi also published (1982) a booklet titled *Soma World* that contains more than 2,000 Soma constructions. The author's address is 19 Yogelsang Place, Flynn, Canberra, Australia.

"A Search for N-Pentacube Prime Boxes." David Klarner in *Journal of Recreational Mathematics*, Vol. 12, No. 4, 1979–1980, pages 252–257.

"Packing Handed Pentacubes." C. J. Bouwkamp in *The Mathematical Gardner*, edited by David Klarner. Prindle, Weber and Schmidt, 1981.

The Soma Cube

"The Soma Cube." Martin Gardner in *The Second Scientific American Book of Mathematical Puzzles and Diversions,* Chapter 6. Simon and Schuster, 1961.

The Solid Pentominoes

Catalog of Solutions of the Rectangular 3 × 4 × 5 Solid Pentomino Problem. C. J. Bouwkamp. Department of Mathematics, Technische Hogeschool Eindhoven, Netherlands, 1967.

"Packing a Rectangular Box with the Twelve Solid Pentominoes." C. J. Bouwkamp in *Journal of Combinatorial Theory,* Vol. 7, 1969, pages 278–280.

"Packing a Box with Y-Pentacubes." C. J. Bouwkamp and D. A. Klarner in *Journal of Recreational Mathematics,* Vol. 3, 1970, pages 10–26. See also Klarner's follow-up letter in the October 1970 issue, page 258.

"A New Solid Pentomino Problem." C. J. Bouwkamp and D. A. Klarner in *Journal of Recreational Mathematics,* Vol. 4, 1971, pages 179–186.

"The F-Pentacube Problem." J. M. M. Verbakel in *Journal of Recreational Mathematics,* Vol. 5, 1972, pages 20–21.

"Solid Pentomino Multiplications." Ad Mank in *Journal of Recreational Mathematics,* Vol. 7, 1974, pages 279–282.

Packing the Steps with Solid Pentominoes. C. J. Bouwkamp. Department of Mathematics, Technische Hogeschool Eindhoven, Netherlands, 1979. Gives the 137 solutions to problem 44 on page 158 of S. W. Golomb's book *Polyominoes.*

CHAPTER FOUR

Bacon's Cipher

Cryptography is a science of deduction and controlled experiment; hypotheses are formed, tested and often discarded. But the residue which passes the test grows and grows until finally there comes a point when the experimenter feels solid ground beneath his feet: his hypotheses cohere, and fragments of sense emerge from their camouflage. The code "breaks." Perhaps this is best defined as the point when the likely leads appear faster than they can be followed up. It is like the initiation of a chain-reaction in atomic physics; once the critical threshold is passed, the reaction propagates itself.

—JOHN CHADWICK, *The Decipherment of Linear B*

It is not hard to understand why philosophers and historians of science are so divided in their opinions about Sir Francis Bacon, the Elizabethan writer, philosopher and Lord Chancellor. On the one hand, his insights into scientific method were primitive and defective. On the other, he had a prophetic vision of science as a vast, collective and systematic enterprise that could provide humanity with undreamed-of knowledge. And knowledge, he insisted, is power. For the first time man would have the power to master nature and control his own destiny.

Although Bacon had little skill in mathematics, he did invent an ingenious cipher system of considerable interest to students of both recreational mathematics and word play. The "biliteral cipher," as Bacon called it, was one of the earliest demonstrations of how easily information can be transmitted by a simple binary code. The system is related to a fascinating combinatorial problem that has practical applications to error-correcting codes. Not least, Bacon's cipher has been responsible for the funniest and most bizarre claims ever propounded by the Baconians — those never-give-up pseudoscholars who still labor mightily to convince the world that Bacon wrote the plays of Shakespeare.

There are hints about the biliteral cipher in Bacon's *Advancement of Learning* (1605), but he did not fully disclose the method until he expanded his brief remarks on ciphers for the later encyclopedic edition of this work in Latin, *De Augmentis Scientiarum* (1623). In Book 6 he repeats his earlier summary of the three virtues every good cipher should have: (1) "Easy and not laborious to write"; (2) "Safe and impossible to decipher"; (3) "If possible, such as not to raise suspicion."

A cipher with the third merit, known as a "concealment cipher," is one in which the very existence of the true cipher text is not suspected. Bacon first explains a whimsical concealment dodge using two cipher alphabets. The genuine message is written with one set of symbols, then a false message is written with a second set. The two ciphers are interwoven to make a single cipher text. If this is intercepted and a translation demanded of the sender, he strikes out the symbols of the true text, explaining that they are what cryptographers today call "nulls," meaningless symbols inserted only to make the cipher harder to break. He then reveals the key to the remaining symbols. Because an intelligible message now emerges, Bacon writes, who would suspect that the apparent nulls actually conceal another message?

"But for avoiding suspicion altogether," Bacon continues, "I will add another contrivance, which I developed myself when I was at Paris in my early youth." The contrivance, the biliteral cipher, is based on a key that assigns to each letter of the alphabet a different permutation of two symbols in groups of five. As Bacon explains, there are 32 such permutations, more than enough for the English alphabet, which in Bacon's day consisted of 24 letters. (*I* and *J* were interchangeable, as were *U* and *V*.) Bacon used *a* and *b* for the two symbols, assigning *aaaaa* to *A*, *aaaab* to *B*, *aaaba* to *C* and so on.

"Nor is it a slight thing which is thus by the way affected," Bacon writes. "For instance we see how thoughts may be communicated at any distance of place by means of any object perceptible either to the eye or ear, provided only that those objects are capable of two differences, as by balls, trumpets, torches, gunshots, and the like." Indeed, the Morse telegraphic code is essentially a biliteral sound cipher, although pauses are used as a kind of third symbol so that no more than four dots and dashes are needed for each letter.

Bacon's plan was to use this cipher for concealing the plaintext (message to be enciphered) in an innocent-looking "cover text." One has only to distinguish between two different ways of printing each letter. A crude method would be to let italicized letters stand for *a* and roman letters for *b*. The word "Bacon," with only the first letter italicized, would represent the permutation *abbbb*, which in

Bacon's alphabet means Q. It is obvious that any cover text, provided it is five times the length of the plaintext, can be printed so that it carries the secret message.

The difference between roman and italicized letters is, of course, too obvious. Bacon proposed using two type fonts that differed in minute ways. Only someone aware of these subtle differences would know how to scan the printing, label each letter *a* or *b*, divide the letters into quintuplets and read the hidden message. Bacon gave two examples of how these fonts could conceal a message. A short Latin cover text meaning "Do not go until I come" deciphers as a message of opposite advice: "Flee." A longer example of how "anything can be written by anything" is a passage Bacon took from a letter of Cicero *[see Figure 24]*. When the letters are labeled *a* and *b* (according to the two fonts), the concealed Latin message (copied from one the Spartans had once sent by a cylindrical ciphering device called a scytale) translates into English as "All is lost. Mindarus is killed. The soldiers want food. We can neither get hence, nor stay longer here."

Elizabethan printing was so crude by modern standards that no two appearances of the same letter on a page, when examined under a strong magnifying glass, are exactly alike. Lead molds were imperfect, type was often damaged, ink dried irregularly on rough and dampened paper, and printers often mixed fonts on the same page. It is not surprising that anyone persuaded that Bacon wrote the plays of Shakespeare would suspect that Bacon might have used his own cipher to state the fact in early folios, perhaps even pepper the pages with other secret revelations.

Elizabethan printing has provided Baconians with a marvelous arena for the unhampered play of unconscious impulses. With a magnifying glass in hand and flexible biliteral rules allowing *a* and *b* forms of each letter to be distinguished in any possible way (and in more than one way for each letter), a clever Baconian can extract from a long passage of Shakespeare's almost any short message he likes. The first appearance of a *T* may be labeled *a* because it has a slightly thinner upright line than other *T*'s; the next *T* may be labeled *a* because it has a tiny curl at the end of the crossbar, and so on. Cipher keys are allowed to vary from passage to passage. If a Baconian is not a mountebank, the secret messages he finds will spring from deep within his subconscious, like the messages spelled on Ouija boards or by automatic handwriting or transmitted by mediums from the Great Beyond.

Strangely enough, the first major effort to decipher Shakespeare's plays did not exploit Bacon's cipher. The flamboyant Populist politician from Minnesota, Ignatius Donnelly, used a different system, even more farfetched, for his 1,000-page crank work *The Great Cryptogram* (1888). (This tome and Don-

Ego omni officio, ac potius pietate erga te.
caeteris satisfacio omnibus: Mihi ipsenunc
quàm satisfacio. Tanta est enim magni=
tudo tuorum erga me meritorum, vt quoni=
am tu, nisi perfectâ re, de me non conquiês=
ti; ego, quia non idem in tuâ causâ efficio,
vitam mihi esse acerbam putem. In cau=
sâ haec sunt: Ammonius Regis legatus
apertè pecuniâ nos oppugnat. Res agitur
per eosdem creditores, per quos, cùm tu ade=
ras, agebatur. Regis causâ, si qui sunt,
qui velint, qui pauci sunt, omnes ad Pompe=
ium rem deferri volunt. Senatus Reli=
gionis calumniam, non religione, sed ma=
leuolentia, et illius Regiae largitionis
inuidiâ comprobat. &c.

Figure 24   A letter of Cicero's in which the two type fonts conceal
a secret war dispatch.

nelly's *Atlantis* and *Ragnarok* form the most impressive set of crackpot works written by an American before 1900.) It remained for Mrs. Elizabeth Wells Gallup (1846–1934), a Michigan teacher and high school principal, to apply Bacon's own cipher with unflagging persistence to Shakespeare's plays, producing the best and most hilarious plaintexts in the history of Baconiana.

Like Donnelly, Mrs. Gallup is a splendid specimen of the intelligent, learned, honest and thoroughly self-deluded crank. Her opus *The Biliteral Cipher of Sir Francis Bacon Discovered in His Works and Deciphered by Mrs. Elizabeth Wells Gallup* (1899) had a shattering impact on fellow Baconians. She found secret messages not only in the Shakespeare folios but also in the writings of Marlowe, Spenser, Burton and other writers whose books she believed had also been written by Bacon. "Queene Elizabeth is my true mother," one message read, "and I am the lawfull heire to the throne. Find the Cipher storie my bookes containe; it tells great secrets, every one of which, if imparted openly, would forfeit my life." Many of the great secrets turned out to be bawdy details of Elizabethan court life.

"Surprise followed surprise," wrote Mrs. Gallup, "as the hidden messages were disclosed, and disappointment as well was not infrequently encountered. Some of the disclosures are of a nature repugnant, in many respects, to my very soul. . . . As a decipherer I had no choice, and I am in no way responsible for the disclosures, except as to the correctness of the transcription."

"Colonel" George Fabyan (the military title was honorary), a wealthy textile manufacturer, became Mrs. Gallup's convert and major benefactor. He brought her to Riverbank Laboratories on his 500-acre estate in Geneva, Ill., where he established a staff of cryptanalysts to work under Mrs. Gallup's supervision. She remained there for 20 years, studying photographic enlargements of Elizabethan manuscripts and trying to teach her bewildered staff how to decipher them.

Ironically, as David Kahn observes in his book *The Codebreakers,* it was at Riverbank that young William F. Friedman was first introduced to the art of code-breaking. Later he became one of the world's greatest cryptanalysts. (It was his team that cracked the Japanese "purple code" of World War II.) While he was at Riverbank, he met and married another of Mrs. Gallup's assistants, Elizabeth Smith. The two eventually became the most illustrious husband-and-wife team in the history of cryptanalysis. Both, I hasten to add, quickly caught on to how Mrs. Gallup was deceiving herself. Indeed, the chapters devoted to Mrs. Gallup in their book *The Shakespearean Ciphers Examined* totally demolish Mrs. Gallup's monumental and pathetic lifetime labors.

Back to mathematical reality. In recent decades mathematicians have developed many ingenious procedures for forming cyclic chains in which all possible

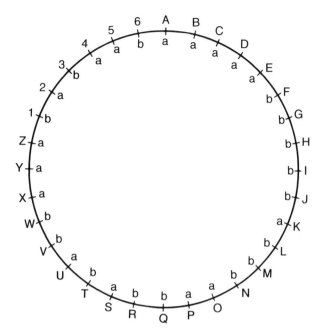

Figure 25 A concise key for a bilateral cipher.

permutations of *n* symbols, taken *k* at a time, are given once only by each set of *k* adjacent symbols. For example, consider the 32-symbol chain

aaaaabbbbbabbbaabbababbaaababaab

If you view the chain as cyclic (end joined to beginning), every group of five adjacent symbols is one of the $2^5 = 32$ permutations of *a* and *b* in sets of five. There are 2,048 ways to construct such a chain, if reversals are considered different. For two symbols the formula giving the number of chains is

$$2^{(2^{k-1}-k)}$$

where *k* is the number of symbols in a group. Any of the 2,048 chains provides a convenient way of recording the key to a biliteral cipher. Simply print the alphabet, with the first six digits appended to bring the number of symbols to 32, in a circle and add the chain of *a*'s and *b*'s inside the circle *[see Figure 25]*. To obtain the permutation for, say, *R*, check the set of five symbols that start at *R* and go clockwise (or the other way if you prefer) around the circle.

The cipher has many unusual applications. A deck of 52 playing cards, for instance, can be arranged so that the colors (or odd and even values, or high and low cards or any other binary division) will encipher a 10-letter word or phrase.

Of course, three-symbol chains provide triliteral ciphers, four symbols provide quadriliteral ciphers (the genetic code!) and so on.

Although it is a defect of Bacon's system that a cipher text must be five times as long as the plaintext, a remarkable merit of the system is that more than one message can be hidden in the same cipher text. One has only to choose letters carefully so that they can be divided into a's and b's in more than one way. Consider, for example,

GkwRt ceUya porrE

Our cipher key will again be the concentric circles in Figure 25, reading clockwise. If a stands for letters whose positions in the alphabet are odd (a, c, e, . . .), and b for even-positioned letters (b, d, f, . . .), the text deciphers as *aaabb aaaaa babba,* which spells CAT. If a refers to a letter in the first half of the alphabet and b to letters in the second half, the same text deciphers as *aabbb aabba bbbba,* which spells DOG. And if a means uppercase and b lowercase, the translation is *abbab bbabb bbbba,* or PIG.

Here is an exercise for readers:

QUZGF MTXYX JLUIY XNEEN WLREW TSNJE

Using the same key as before, can you determine three ways of bifurcating the alphabet so that the above cipher text can be translated in three ways, each giving a six-letter last name of a famous mathematician? (Hints: The three divisions have to do with the name of a poet, legs and topology.)

Although Bacon himself did not make the metaphor explicit, his cipher may be taken as symbolic of the curious way he viewed scientific knowledge. It is an attitude still held today by many philosophers and scientists. Bacon did not believe that the laws of science were infinite in number. Like his fellow Anglicans, he was convinced that God had created a natural world that was sharply cut off from the supernatural. In *this* world a finite number of simple principles combine, like the variables of an n-literal cipher, to form all the laws of nature.

The 19th-century English logician John Venn made this point in his *Empirical Logic* (page 357), where he described Bacon's position as an "*alphabetical view of the Universe, in its extremest form.* . . . We find [the universe] all broken up, partitioned, and duly labeled in every direction; so that, enormously great as is the possible number of combinations which these elements can produce, they are nevertheless *finite* in number, and will therefore yield up their secrets to plodding patience when it is supplied with proper rules."

Science, to pursue the metaphor, is one stupendous task of cryptanalysis. Bacon was persuaded that eventually, and not far in the future either, all the ciphers would be broken and mankind would know not all truth by any means, but all the basic natural laws. The future of science would then be merely a filling in of details and the exploitation of laws by new inventions.

Although few scientists today would venture such a prediction, more limited Baconian sentiments are often expressed with reference to a particular science. Nigel Calder, in his vivid survey of the new astronomy, *Violent Universe* (Viking, 1969), suggests that our century may turn out to be unique in the history of astronomy as the century in which astronomers first became "know-alls," omniscient in the sense of having mapped the fundamental outlines of the entire cosmos. "Or," Calder adds, "will our descendants smirk about our ideas as we do about those of our ancestors?"

Who can be sure, even with reference to a single science, whether in the long run (whatever that means) Bacon will be proved right or wrong? We *can* say that at the moment nature appears to be far shaggier and more complicated than the Lord Chancellor suspected. There are ciphers within ciphers within ciphers, and there is not a clue in sight about whether any of these regresses has an end.

ANSWERS

The three translations of the Baconian cipher are Fermat, Galois, Newton. The three biliteral keys respectively are

1.   Any letter in WILLIAM SHAKESPEARE is *a*; all others are *b*.

2.   Any letter with one or more legs when printed as a capital is *a* (A, F, H, I, K, M, N, P, Q, R, T, X, Y). No-leg letters are *b*.

3.   Any letter that in simplest capital form is topologically equivalent to a line segment is *a* (C, I, L, M, N, S, U, V, W, Z). All others are *b*.

ADDENDUM

I received a fascinating letter from Marguerite Gerstell, then an instructor at the Florida Institute of Technology in Jensen Beach. Using the same circular key that I used for my puzzle, she encoded the names of *five* eminent mathematicians in the following cipher text:

HUUSN IUUII YPDAW WVALP EZRWZ TISOS

"Four of them are easy to find," she wrote. "Anyway, a smart gal can help you with the fifth."

Here is how she did it:

1.   NAPIER is encoded by replacing all vowels (including Y) in the cipher text with *a* and all other letters with *b*.

2.   EUCLID is encoded by replacing with *a* all letters whose ordinals (position in the alphabet) are multiples of a square greater than 1.

3.   KUMMER is encoded by substituting *a* for letters that are among the first 15 of the alphabet.

4.   CAUCHY is encoded by replacing left–right symmetrical letters by *a*.

5.   CANTOR is encoded by substituting *a* for letters *not* in the phrase "anyway a smart gal."

A surprising thought occurred to Gerstell. Why not use the name itself as a basis for distinguishing *a* and *b*? She sent four examples of cipher texts, each concealing the names of three mathematicians, by using this curious self-reference technique. Here is one of them:

ZYMWL EIGAI UMBOI JULRY MYFGA IXYZM LOSUL

The three names are Zermelo, Galileo and Fourier. In each case the name is encoded by letting *a* stand for letters *in* the name being concealed. As Gerstell pointed out, it is not easy to accomplish this with more than three names. It would be an interesting challenge, she wrote, to try to maximize the number of names that could be simultaneously encoded in this way.

Gerstell's cipher texts all use the same cyclic chain that I suggested for a biliteral cipher. Such chains are now known as de Bruijn sequences, after the Dutch mathematician N. G. de Bruijn. For a fascinating history of such chains see "Memory Wheels," by Sherman K. Stein, in the second edition of his *Mathematics: The Man-Made Universe* (Freeman, 1969). In recent years mathematically minded magicians have invented a variety of bewildering card tricks based on de Bruijn sequences. References to where you can find some of them are in the answer section of Chapter 12 in my *Magic Numbers of Dr. Matrix* (Prometheus, 1985). For a recent article, with a good bibliography, on de Bruijn sequences see "De Bruijn Sequences—A Modern Example of the Interaction of Discrete Mathematics and Computer Science," by Anthony Ralston in *Mathematics Magazine,* Vol. 55, 1982, pages 131–143.

Someone ought to write a book about the sad life of Mrs. Gallup. Little seems to be on record about her. Apparently she taught at various public schools in Michigan (at Wayne, Flint, Fenton and Holly) and was a principal of the Holly high school. Friedman says she died in 1934, but an obit in the British periodical *Baconia* (October, 1935, page 106), called to my attention by David Shulman, gives the date of her death as April 1933 and her age as 87. She was born February 4, 1846, near Waterville, N.Y., educated at State Normal College of Michigan and was later a graduate student at the University of Marburg and the Sorbonne. I have been unable to determine what subject she taught or who Mr. Gallup was.

All her tomes were published by Howard Publishing Company, Detroit, which I take to be her own company. The first edition of her opus (1899) was a mere 246 pages, but the second edition (1900) expanded it to 480 pages. The third edition (1901) is even larger — two volumes. In 1902 she issued a booklet titled *Bi-literal Cipher of Francis Bacon: Replies to Criticisms. Concerning the Bi-literal Cipher of Francis Bacon, Discovered in His Works: Pros and Cons of the Controversy* was a 1910 book of 229 pages. She also published (1901) a 147-page work titled *The Tragedy of Anne Boleyn: A Drama in Cipher Found in the Works of Sir Francis Bacon.*

A bibliography of articles about Mrs. Gallup's obsessions would run to many pages. Here are the few references I was able to track down.

"Mrs. Gallup's Cipher." *Blackwood's Magazine,* Vol. 171, 1902, pages 269 – 267.

"Mrs. Gallup and Francis Bacon." Andrew Lang in *The Monthly Review,* Vol. 2, 1902, pages 146 – 162.

"Mrs. Gallup's Bad History." Robert S. Rait in *Fortnightly Review,* Vol. 77, 1902, pages 328 – 334.

Studies in the Bi-literal Cipher of Francis Bacon. Gertrude Horsford Fiske. J. W. Luce, 1913.

"The Encyclopedia Britannica and Mrs. Gallup." B. Wright in *Baconia,* No. 132, 1949, pages 154 – 160.

A picture of Mrs. Gallup can be found in Friedman's book, cited earlier, and a different photograph appears in all her books.

Georg Cantor, by the way, the genius who founded modern set theory, was a passionate believer in the Bacon – Shakespeare theory. During his later years of manic depression, when he was dabbling in theosophy and other occult matters, he wasted enormous amounts of time trying to prove the theory, lecturing on the topic and writing many articles. Cantor believed that his set theory had been directly inspired by God and was therefore flawless. His biblical studies convinced him that Jesus was the natural son of Joseph of Arimathea, and he wrote the pamphlet *Ex Oriente Lux* to prove it. (See "Georg Cantor's Creation of Transfinite Set Theory: Personality and Psychology in the History of Mathematics," by Joseph W. Dauben in *Annals of the New York Academy of Sciences,* Vol. 321, 1979, pages 27 – 44, a volume titled *Papers in Mathematics,* edited by Paul Meyer.)

I closed my column by expressing doubts that science was near discovering that everything in physics could be explained, as Bacon suggested, by a finite set of laws. At the moment this hope has sprung up again among many top physicists, who believe they are on the verge of constructing a grand unified-field theory that will cover all the forces of nature and explain why all the particles are just what they are. See my review of two recent books expressing this euphoria: "Physics: The End of the Road?", in *The New York Review of Books,* June 13, 1985, pages 31 – 34.

BIBLIOGRAPHY

The Philosophy of Francis Bacon. Fulton H. Anderson. University of Chicago Press, 1948.

The Shakespearean Ciphers Examined. William F. and Elizabeth S. Friedman. Cambridge University Press, 1957.

The Codebreakers. David Kahn. Macmillan, 1967.

"Origins of the Binary Code." F. G. Heath in *Scientific American,* August, 1972.

CHAPTER FIVE

Doughnuts:
Linked and Knotted

As you ramble on through life, brother,
 Whatever be your goal,
Keep your eye upon the doughnut
 And not upon the hole!

 —ANON.

A torus is a doughnut-shaped surface generated by rotating a circle around an axis that lies on the plane of the circle but does not intersect the circle. Small circles, called meridians, can be drawn around the torus with radii equal to that of the generating circle. Circles of varying radii that go around the hole or center of the torus on parallel planes are called parallels [*see Figure 26*]. Both meridians and parallels on a torus are infinite in number. There are two other less obvious infinite sets of "oblique" circles with radii equal to the distance from the center of the generating circle to the center of the torus's hole. Can you find them? Members of one set do not intersect one another, whereas any member of one set twice intersects any member of the other.

To a topologist, concerned only with properties that do not alter when a figure is elastically deformed, a torus is topologically equivalent to the surface of such objects as a ring, a bagel, a life preserver, a button with one hole, a coffee cup, a soda straw, a rubber band, a sphere with one handle, a cube with one hole through it and so on. Think of these surfaces as a thin membrane that can be stretched or compressed as much as one wishes. Each can be deformed until it

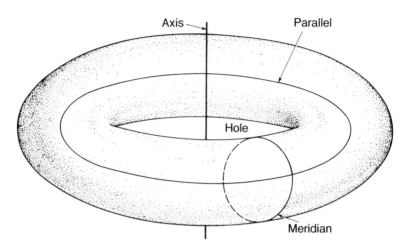

Figure 26   The torus.

becomes a perfect toroidal surface. In what follows, "torus" will mean any surface topologically equivalent to a torus.

A common misunderstanding about topology is the belief that a rubber model of a surface can always be deformed in three-dimensional space to make any topologically equivalent model. This often is not the case. A Möbius strip, for example, has a handedness in 3-space that cannot be altered by twisting and stretching. Handedness is an extrinsic property it acquires only when embedded in 3-space. Intrinsically it has no handedness. A 4-space creature could pick up a left-handed strip, turn it over in 4-space and drop it back in our space as a right-handed model.

A similar dichotomy applies to knots in closed curves. Tie a single overhand (or trefoil) knot in a piece of rope and join the ends. The surface of the rope is equivalent to a knotted torus. It has a handedness, and no amount of fiddling with the rope can change the parity. Intrinsically the rope is not even knotted. A 4-space creature could take from us an unknotted closed piece of rope and, without cutting it, return it to us as knotted in either left or right form. All the properties of knots are extrinsic properties of toruses (or, if you prefer, one-dimensional curves that may be thought of as toruses whose meridians have shrunk to points) that are embedded in 3-space.

It is not always easy to decide intuitively if a given surface in 3-space can be elastically deformed to a different but topologically equivalent surface. A striking instance, discussed more than 20 years ago [see "Topology," by Albert W. Tucker and Herbert S. Bailey, Jr., in *Scientific American*, January, 1950], concerns a rubber torus with a hole in its surface. Can it be turned inside out to

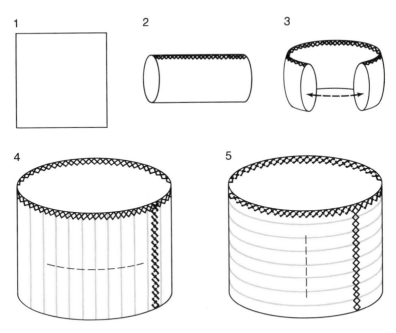

Figure 27   Reversible cloth torus.

make a torus of identical shape? The answer is yes. It is hard to do with a rubber model (such as an inner tube), but a model made of wool reverses readily. Stephen Barr, in his *Second Miscellany of Puzzles* (Macmillan, 1969), recommends making it from a square piece of cloth. Fold the cloth in half and sew together opposite edges to make a tube. Now sew the ends of the tube together to make a torus that is square shaped when flattened. For ease in reversing, the surface hole is a slot cut in the outer layer of cloth *[shown by the broken line in Figure 27]*.

After the cloth torus is turned inside out, it is exactly the same shape as before, except that what were formerly meridians have become parallels, and vice versa. To make the switch visible, sew or ink on the model a meridian of one color and a parallel of another so that both colors are visible from either side of the cloth. In 1958 Mrs. Eunice Hakala sent me a model she had made by cutting off the ribbed top of a sock and joining the tube's ends. The ribbing provides a neat set of parallels that turn into meridians after the torus is reversed.

Let us complicate matters by considering a torus tied in a trefoil knot. If we ignore handedness, there are only two such toruses: one with an external knot and one with an internal knot *[see Figure 28 a,b]*. A way to visualize the

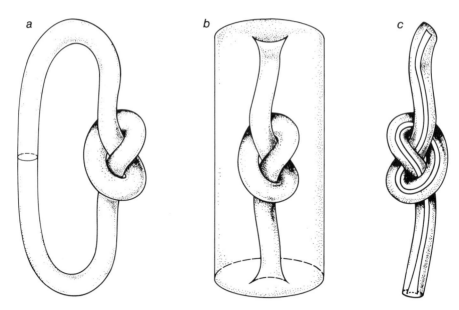

Figure 28 Torus with outside knot (*a*), inside knot (*b*) and pseudoknots (*c*).

internally knotted torus is to imagine that the externally knotted torus on the left is sliced open along a meridian outside the knot. One end is turned back, as though reversing a sock; then the tube is expanded and drawn over the entire knot, and its ends are joined once more. Or imagine a solid wood cube with a hole bored through it that, instead of going straight, ties a knot before it emerges on the opposite side. The surface of such a cube is topologically equivalent to an internally knotted torus.

You might suppose that a torus could be simultaneously knotted externally and internally, but it can't be done. One kind of torus seems to have both an outside and an inside knot [*see Figure 28c*]. Actually both knots are humbugs. Untying the outer knot simultaneously unties the inner one, proving that the model is topologically the same as an unknotted torus — its hole elongated like the hole of a garden hose.

Although an outside-knotted torus is intrinsically identical with an inside-knotted one, it is not possible to deform one to the other when it is embedded in 3-space. If there is a hole in the side of an outside-knotted torus, can the torus be reversed in 3-space to put the knot inside? In the answer section I shall show how R. H. Bing, a topologist at the University of Wisconsin, answers this question with a simple sketch.

A similar but harder problem is solved by Bing in his paper "Mapping a 3-Sphere onto a Homotopy 3-Sphere," in *Topology Seminar, Wisconsin, 1965*, edited by Bing and R. J. Bean (Princeton University Press, 1966). Imagine a cube with two straight holes [*see Figure 29a*]. Its surface is topologically the same as a two-hole doughnut. We can also have a cube with two holes, one straight,

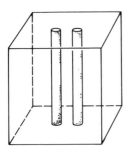
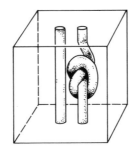
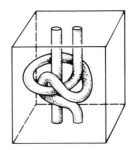

Figure 29   Three varieties of a two-hole torus.

one knotted *[Figure 29b]*. It is not possible in 3-space to deform the second cube so that the knot dissolves and the model looks like the first one. A third cube has one straight hole and one knotted hole with the knot around the straight hole *[Figure 29c]*. Can this cube be elastically deformed until it becomes the first model? It is hard to believe, but the answer is yes. Bing's proof is so elegant and simple that the diagrams for it are almost self-explanatory *[see Figure 30]*. In elastic deformation a hole can be moved any distance over a surface without altering the surface's topology. As the hole moves, the surface merely stretches in back and shrinks in front. In Bing's proof the knotted tube is drawn as a single

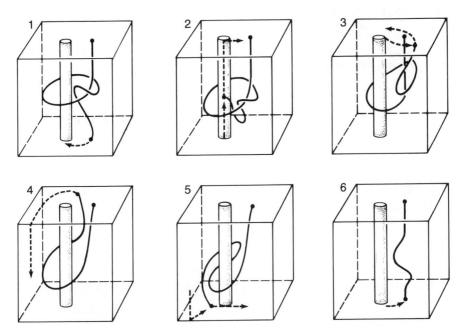

Figure 30   R. H. Bing's proof.

line to make the proof easier to follow. The hole, at the base of this tube is moved over the cube's surface, as indicated by the arrows, dragging the tube along with it. It goes left to the base of the other tube, climbs that tube's side, moves to the right across the top of the cube, circles its top hole counterclockwise, continues left around the other hole, over the cube's front edge, down the front face, around the lower edge to the cube's bottom face and then across that face to the position it formerly occupied. It is easy to see that the tube attached to this hole has been untied. Naturally the procedure is reversible. If you had a sufficiently pliable doughnut surface with two holes, you could manipulate it until one hole became a knot tied around the other.

Topologists worried for decades about whether two separate knots side by side on a closed rope could cancel each other; that is, could the rope be manipulated until both knots dissolved? No pair of canceling knots had been found, but proving the impossibility of such a pair was another matter. It was not even possible to show that two trefoil knots of opposite handedness could not cancel. Proofs of the general case were not found until the early 1950's. One way of proving it is explained by Ralph H. Fox in "A Quick Trip through Knot Theory," in *Topology of 3-Manifolds and Related Topics,* edited by M. K. Fort, Jr. (Prentice-Hall, 1963). It is a *reductio ad absurdum* proof that unfortunately involves the sophisticated concept of an infinity of knots on a closed curve and certain assumptions about infinite sets that must be carefully specified to make the proof rigorous.

When John Horton Conway, the University of Cambridge mathematician, was in high school, he hit on a simpler proof that completely avoids infinite sets of knots. Later he learned that essentially the same proof had been formulated earlier, but I have not been able to determine by whom. Here is Conway's version as he explained it years ago in a letter. It is a marvelous example of how a knotted torus can play an unexpected role in proving a fundamental theorem of modern knot theory.

Conway's proof, like the one for the infinite knots, is a *reductio ad absurdum.* We begin by imagining that a closed string passes through the opposite walls of a room [see Figure 31]. Since we shall be concerned only with what happens inside the room, we can forget about the string outside and regard it as being attached to the side walls. On the string are knots A and B. Each is assumed to be genuine in the sense that it cannot be removed by manipulating the string if it is the only knot on the string. It also is assumed that the two knots will cancel each other when both are on the same closed curve. The proof applies to pairs of knots of any kind whatever, but here we show the knots as simple trefoils of opposite parity. If the knots can cancel, it means that the string can be manipu-

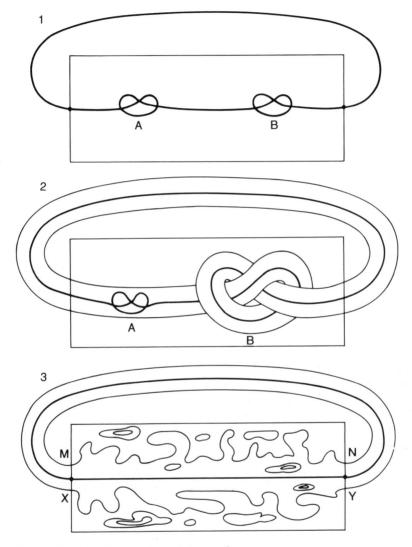

Figure 31   John Horton Conway's proof.

lated until it stretches straight from wall to wall. Think of the string as being elastic to provide all the needed slack for such an operation. In the center figure we introduce an elastic torus around the string. Note that the tube "swallows" knot A but "circumnavigates" knot B (Conway's terminology). Any parallel drawn on this tube, on the section between the walls, obviously must be knotted in the same way as knot B. Indeed, it can be shown that any line on the tube's surface, stretching from wall to wall and never crossing itself at any spot on the tube's surface, will be knotted like knot *B*.

"Now," writes Conway, "comes the crunch." Perform on the string the operation that we assumed would dissolve both knots. This can be done without breaking the tube. Because the string is never allowed to pass through itself during the deformation, we can always push the tube's wall aside if it gets in the way. The third drawing in Figure 31 shows the final result. The string is unknotted. The tube may have reached a horribly complicated shape impossible to draw. Consider a vertical plane passing through the straight string and cutting the twisted tube. We can suppose that the tube's cross section will look something like what is shown with the possibility of various "islands," but there will necessarily be two lines, *XY* and *MN*, from wall to wall that do not cross themselves at any point on the vertical plane. Each line will be unknotted. Moreover, each line also is a curve that does not cross itself on the tube's surface. As we have seen, all such lines were (before the deformation) knotted like knot *B*. The deformation has therefore removed a knot equivalent to knot *B* from each of these two lines. Therefore knot *B*, alone on a line, can be removed by manipulating that line. But knot *B*, by definition, is a genuine knot that cannot be so removed. We have contradicted an assumption. If two knots on a string can cancel, neither knot (since the same proof can be applied to knot *A*) can be genuine. Both must really have been pseudoknots.

Although a one-hole torus can be embedded in 3-space in only three ways (outside knot, inside knot, no knot), a two-hole torus has so many bizarre forms that the number is, I believe, not yet known. In some cases it can be reduced to a simpler form by deformation. For example, a tube-through-hole is equivalent to an ordinary two-hole doughnut *[see Figure 32]*, but what about the other two figures *[b and c]*? They are among several dozen monstrosities sketched by Piet Hein in a moment of meditation on two-hole toruses. In *b* an inside knot goes through an outside one, and in *c* an outside knot goes through a hole. Is it possible, by deformation, to dissolve the inside knot of *b* and the outside knot of *c*?

With more complicated pairs of two-holers embedded in 3-space, proofs that one can be deformed to the other are not so easy. As one of Piet Hein's "grooks" puts it:

> There are doughnuts and doughnuts
> with knots and with no knots
> and many a doughnut
> so nuts that we know not.

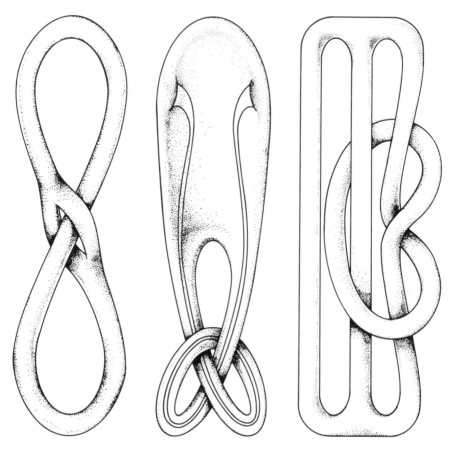

Figure 32   Two-hole toruses.

Here are three more toroidally knotty questions.

1.   How many closed curves can be drawn on a torus, each a trefoil knot of the same handedness, so that no two curves cross each other at any point?

2.   If two closed curves are drawn on a torus so that each forms a trefoil knot but the knots are of opposite parity, what is the minimum number of points at which the two curves will intersect each other?

3.   Show how to cut a solid two-hole doughnut with one slice of a knife so that the result is a solid outside-knotted torus. The "slice" is not, of course, planar. More technically, show how to remove from a two-hole doughnut a section topologically equivalent to a disk so that what remains is a solid knotted torus. (This amusing result was discovered by John Stallings in 1957 and communicated to me by James Stasheff.)

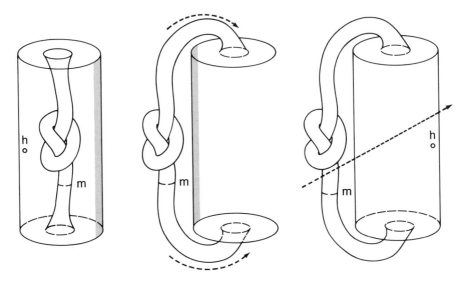

Figure 33   Solution to torus-reversed problem.

ANSWERS

R. H. Bing shows how an internally knotted torus can be reversed through a hole to produce an externally knotted torus *[see Figure 33]*. A small hole, *h*, is enlarged to cover almost the entire side of the cylinder, leaving only the shaded strip on the right. The top and bottom disks of the cylinder are flipped over, and the hole is shrunk to its original size.

As in reversing the unknotted torus through a hole, the deformation interchanges meridians and parallels. You might not at first think so because the circle, *m*, appears the same in all three pictures. The fact is, however, that initially it is a parallel circling the torus's elongated hole, whereas after the reversal it is a meridian. Moreover, after the reversal the torus's original hole is no longer through the knotted tube, which is now closed at both ends. As indicated by the arrow, the hole is now surrounded by the knotted tube.

Piet Hein's two-hole torus, with an internal knot passing through an external one, is easily shown to be the same as a two-holer with only an external knot. Simply slide one end of the inside knot around the outside knot (in the manner explained earlier) and back to its starting point. This unties the internal knot. Piet Hein's two-holer, with the external knot going through a hole, can be unknotted by the deformation shown in Figure 34.

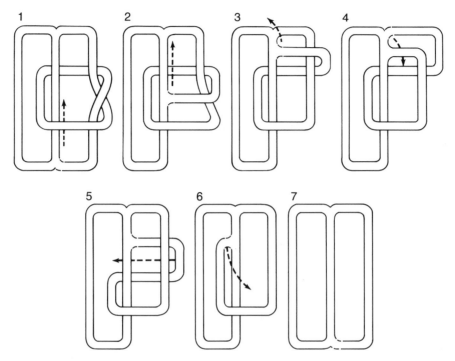

Figure 34   Unknotting a two-hole torus.

Answers to the final three toroidal questions are as follows:

1.   An infinity of noncrossing closed curves, each knotted with the same handedness, can be drawn on a torus *[see Figure 35 top]*. If a torus surface is cut along any of these curves, the result is a two-sided, knotted band.

2.   Two closed curves on a torus, knotted with opposite handedness, will intersect each other at least 12 times.

3.   A rotating slice through a solid two-hole doughnut is used to produce a solid that is topologically equivalent to a solid, knotted torus *[see Figure 35, bottom]*. Think of a short blade as moving downward and rotating one and a half turns as it descends. If the blade does not turn at all, two solid toruses result. A half-turn produces one solid, unknotted torus. One turn produces two solid, unknotted, linked toruses. Readers may enjoy investigating the general case of *n* half-turns.

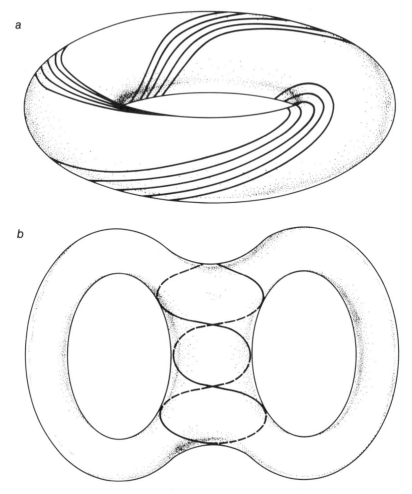

Figure 35   Knotted, nonintersecting curves on a torus (*a*) and rotating slice through a two-hole torus (*b*)

ADDENDUM

In studying the properties of topological surfaces, one must always keep in mind the distinction between intrinsic properties, independent of the space in which the surface is embedded, and properties that arise from the embedding. The "complement" of a surface consists of all the points in the embedding space that are not in the surface. For example, a torus with no knot, one with an outside knot and one with an inside knot all have identical intrinsic properties. No two have topologically identical complements; hence, no two are equivalent in their extrinsic topological properties.

John Stillwell, a mathematician at Monash University, Australia, sent several fascinating letters, in which he showed how an unknotted torus with any number of holes—such toruses are equivalent to the sufaces of spheres with

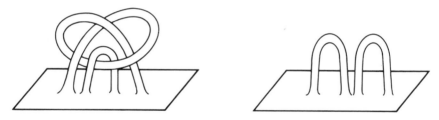

Figure 36 The surface on the left can be continuously deformed to the surface on the right.

handles — could be turned inside out through a surface hole. He was not sure if a knotted torus, even with only one hole, can be turned inside out through a hole in its surface. I leave this as a problem for the reader.

Stillwell also posed the following question. Suppose two ordinary doughnut surfaces are linked, and one has a hole in its surface. Can the torus with the surface hole "swallow" the other torus so that at the finish the eaten torus is completely inside the cannibal? The answer is yes. I gave this problem in my April 1977 column in *Scientific American;* the answer appeared the following month.

Many beautiful, counterintuitive problems involving links and knots in toruses have been published. See Rolfsen's book, cited in the bibliography, especially the startling problem on page 95, where he shows that the surface on the left of Figure 36 is topologically equivalent to the surface shown on the right. For other curious equivalences of this sort see Herbert Taylor's torus problem in my *Scientific American* column for December 1979, and "The Toroids of Dr. Klonefake," Problem 9, in my *Science Fiction Puzzle Tales* (Clarkson Potter, 1981).

BIBLIOGRAPHY

"Elementary Point Set Topology." R. H. Bing in *The American Mathematical Monthly,* Part II, Vol. 67, 1960, special supplement to Part I.

Intuitive Concepts in Elementary Topology. Bradford Henry Arnold. Prentice-Hall, 1962.

Introduction to Knot Theory. Richard H. Crowell and Ralph H. Fox. Springer-Verlag, 1963.

First Concepts of Topology. W. G. Chinn and N. E. Steenrod. Random House New Mathematical Library, 1966.

Knots and Links. Dale Rolfsen. Publish or Perish, 1976.

CHAPTER SIX

The Tour of the Arrows and Other Problems

1. THE TOUR OF THE ARROWS

Sketch a large 4-by-4 checkerboard on a sheet of paper, obtain 16 paper matches, and you are set to work on this new solitaire puzzle. The matches represent arrows that point in the direction of the match head. Put a single spot on both sides of one match, two spots on both sides of eight matches and three spots on both sides of seven matches. When a match is placed on a square of the board pointing north, south, east or west, the single spot means it points to the immediately adjacent cell, two spots mean it points to the second cell and three spots mean it points to the third cell.

Seven matches can be placed to map a closed tour [see Figure 37]. Start at any match of the seven and place your finger on the cell to which it points. The arrow on that cell gives the next "move." Follow the arrows until you return (in seven moves) to where you started. The problem is to place all 16 matches, one to a cell so that they map a closed tour that visits every cell. There are just two solutions, not counting rotations and reflections.

The tour will have a length of $1 + (2 \times 8) + (3 \times 7) = 38$. It is not hard to prove that this is the longest closed tour that can be made on the board by using any combination of the three types of arrows. Brian R. Barwell, a British engineer who introduced the problem in the *Journal of Recreational Mathematics* (October, 1969), found that only one other maximum-length tour is possible. It requires six 3-arrows, ten 2-arrows and no 1-arrow. Readers are invited to search for all three patterns.

The arrows are, of course, merely a convenient way to map a maximum-length, closed tour by a chess rook, which lands on each cell exactly once. (Queen tours of this type are less interesting because there are so many of them;

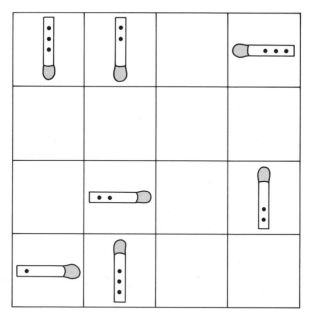

Figure 37 A closed arrow tour

bishop tours cannot close and cannot visit all cells; and knight tours cannot vary in length.) The 2-by-2 field is trivial, and the 3-by-3 is easily analyzed. (Its maximum tour has a length of 14.) As far as I know, the 5-by-5 and all higher squares have yet to be investigated.

2. FIVE COUPLES

My wife and I recently attended a party at which there were four other married couples. Various handshakes took place. No one shook hands with himself (or herself) or with his (or her) spouse, and no one shook hands with the same person more than once.

After all the handshakes were over, I asked each person, including my wife, how many hands he (or she) had shaken. To my surprise each gave a different answer. How many hands did my wife shake? (From Lars Bertil Owe of Lund, Sweden.)

3. SQUARE-TRIANGLE POLYGONS

An unlimited number of cardboard squares and equilateral triangles, each with unit sides, are assumed to be available. With these pieces it is easy to form convex polygons with from 3 to 10 sides *[see Figure 38]*. Can you make an

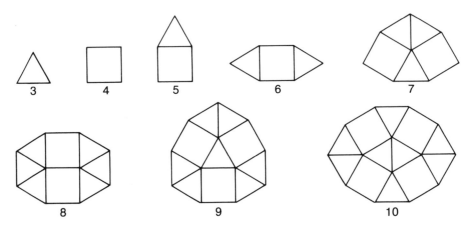

Figure 38   Convex polygons with from 3 to 10 sides

11-sided convex polygon with the pieces? And what is the largest number of sides a convex polygon formed by the pieces can have?

4. TEN STATEMENTS

Evaluate each of the 10 statements as to its truth or falsity:

1. Exactly one statement on this list is false.

2. Exactly two statements on this list are false.

3. Exactly three statements on this list are false.

4. Exactly four statements on this list are false.

5. Exactly five statements on this list are false.

6. Exactly six statements on this list are false.

7. Exactly seven statements on this list are false.

8. Exactly eight statements on this list are false.

9. Exactly nine statements on this list are false.

10. Exactly ten statements on this list are false.

5. PENTOMINO FARMS

Victor G. Feser of Saint Louis University has proposed four maximum-area problems, each using the full set of 12 pentominoes. Three have been solved, and the fourth is probably solved.

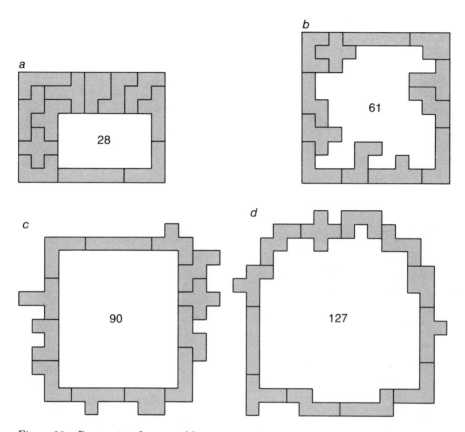

Figure 39 Pentomino fence problems

1. Form a rectangular "fence" around the largest rectangular field. The 4-by-7 has been proved maximum [*see Figure 39a*].

2. Form a rectangular fence around the largest field of any shape. The maximum is 61 unit squares [*see Figure 39b*].

3. Form a fence of any shape around the largest rectangular field. The 9-by-10 is maximum [*see Figure 39c*].

4. Form a fence of any shape around the largest field of any shape. (As in the preceding problems, the fence must be at least one unit thick at all points.) This is the most difficult of the four. In Figure 39d you see a solution of 127 squares. This was believed to be maximum until Donald E. Knuth, the Stanford computer scientist, recently raised it to 128. Knuth has an informal proof that 128 cannot be exceeded. Readers will find it a pleasant and difficult task to find a 128 solution.

6. THE UNEVEN FLOOR

A kitchen has an uneven floor. There are no "steps," but the continuous random waviness of the linoleum is such that when one tries to place on it a small square table with four legs, one leg is usually off the floor, causing the table to wobble. If one does not mind the table top being on a slant, is it always possible to find a place where all four legs are firmly on the floor? Or can a floor wave in such a way that no such spot is available? The problem can be answered by a simple, elegant proof.

7. THE CHICKEN-WIRE TRICK

This strange parlor trick comes from Tan Hock Chuan, a Chinese professional magician who lives in Singapore. He described it in a letter to Johnnie Murray, an amateur conjuror of Portland, Maine, who passed it on to me.

A blank sheet of paper about eight by five inches (half a sheet of typewriter paper works nicely) is initialed by an onlooker so that later it can be identified. The magician holds it behind his back (or under a table) for about 30 seconds. When he brings it back into view, it is covered with creases that form a regular hexagonal tessellation [see Figure 40]. How is it done? The performer is usually accused of pressing it against a piece of chicken wire, but the creasing actually is done without using anything except the hands.

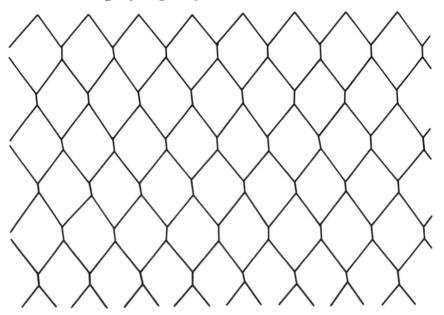

Figure 40   Chicken-wire folds

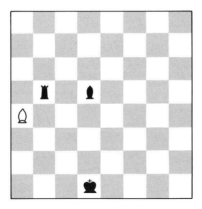

Figure 41   Where was the white king?

8. WHERE WAS THE KING?

The philosopher-mathematician-logician Raymond Smullyan invented this elegant chess problem when he was a student at the University of Chicago in 1957. He showed it to his friend William Browder, now a distinguished mathematician at the university, who passed it on to his father, Earl Browder, former head of the Communist Party in the U.S. and an ardent chess player. The father sent it to the *Manchester Guardian,* where it was inadvertently published without mentioning Smullyan. A later issue gave proper credit for the problem, and other retrograde problems by Smullyan ran in subsequent issues.

A retrograde chess problem is one that can be solved only by deducing the moves that precede the position shown. In this case we see in Figure 41 a position in a legal game just after the white king has been knocked off the board. Where was the king standing? and what was White's last move?

9. POLYPOWERS

By convention, the value of a ladder of exponents such as

$$2^{2^{2^{2^{2}}}}$$

is computed by starting at the top and working down. The highest pair equals 4, then $2^4 = 16$, and $2^{16} = 65{,}536$. How large is $2^{65{,}536}$? A few years ago Geoffrey W. Hoffmann of West Germany sent me a computer printout of this number. It starts 20035 . . . and has 19,729 digits. Adding another 2 to the ladder gives a number that will never be calculated because the answer, as Hoffmann put it, would require the age of the universe in computer time and the space of the universe to hold the printout.

Even a ladder as short as three 9's is $9^{387{,}420{,}489}$, a number of more than 360 million digits. In 1933 S. Skewes published a paper in which he showed that if

$\pi(x)$ is the number of primes less than x, and $\mathrm{li}(x)$ is the logarithmic integral function, then $\pi(x) - \mathrm{li}(x)$ is positive for some x less than

$$10^{10^{10^{34}}}$$

an integer said to be the largest known to play a role in a nontrivial theorem.

In 1971 Aristid V. Grosse, one of the pioneer atomic chemists at Columbia University in 1940 (he is now president of Germantown Laboratories, Inc., affiliated with the Franklin Institute), began an investigation of exponential ladders of identical numbers that are calculated in the opposite direction (up) and their relations to down ladders. He coined the term "polypowers" for ladders of both types. Ladders of two x's are called "dipowers," of three x's "tripowers" and so on, according to the Greek prefixes. The value of x can be rational or irrational, transcendental, complex or entirely imaginary. In most cases the polypowers are single valued, continuous and differentiable. Since 1 to any polypower of 1 is 1, all these functions and their derivatives, when graphed against x, cross one another at $x = 1$, and their values at 0 are the limits as x approaches 0. Grosse's notes, which already fill many volumes, lead into a lush jungle of unusual theorems as well as new classes of numbers.

Up and down dipowers obviously are identical, but for all higher polypowers the two directions give different numbers. The triplet of 9's, for example, when calculated upward is a number of only 77 digits. Except for the triplet of 2's, going "up all the way" on a ladder of identical integers gives the minimum number, and "down all the way" gives the maximum. In what follows, the arrows indicate these maximum and minimum numbers.

What happens when up and down ladders of different lengths are equated? If an up triplet of x's equals a down triplet of x's, $x = 2$. (We exclude $x = 1$ as being trivial.) Each additional x on the up ladder increases the value of x by 1. If three down x's equal four up x's, $x = 3$; if three down equals five up, $x = 4$ and so on.

As an introduction to polypowers, readers are asked to solve the three equations below, which begin a series with down tetrapowers on the left:

$$\overset{\nearrow x^{x^{x}}}{_{x^{x}}} = \overset{\ x^{x^{x}}}{_{x^{x}\nearrow}}$$
$$\overset{\nearrow x^{x}}{_{x^{x}}} = \overset{\ x^{x^{x}}}{_{x^{x}\nearrow}}$$
$$\overset{\nearrow x^{x}}{_{x^{x}}} = \overset{\ x^{x}}{_{x^{x^{x}}\nearrow}}$$

Readers may enjoy investigating ladders of fractional x's, reciprocals of x and more exotic forms. Grosse has also developed the concept of a perfect polypower, that is, x to the xth power (up or down) an x number of times. (Example:

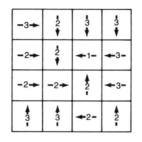

Figure 42 Answer to arrow tours

π to the πth power, π times up, is 588,916,326+.) The reverse operation to polypowers he calls "polyroots." Have these fields been investigated before? In spite of considerable effort, neither he nor I have uncovered references.

ANSWERS

1. The three ways of forming maximum-length arrow tours on the 4-by-4 field are shown in Figure 42.

Edward N. Peters, on the faculty of the University of Rochester Medical School, discovered a general procedure for constructing maximum-length rook tours on square boards of any size. See his paper "Rooks Roaming Round Regular Rectangles," in *Journal of Recreational Mathematics*, Vol. 6, 1973, pages 169–173.

Frederick Hartmann of Rolling Hills Estate, Calif., extended the analysis to nonsquare rectangular boards, but so far as I know, his results remain unpublished. When the board is $n \times 1$, it reduces to the "worst-route" problem for a postman delivering mail to a row of n houses (see my *Sixth Book of Mathematical Games from Scientific American*, W. H. Freeman, 1971, Chapter 23). Maximum-length rook tours on these linear boards are unique from $n = 1$ through 4, then increase in number steadily as n exceeds 4. For $n = 7$, for example, there are 18 such tours.

Hartmann gave an algorithm for constructing at least one maximum-length tour on any rectangular board. If m and n are the lengths of the sides, with m equal to or greater than n, and C is obtained from the table shown in Figure 43, the formula for the length of the tour is

$$\frac{n(3m^2 + n^2 - 10)}{6} + C$$

For square boards of side n the formula reduces to

$$\frac{2n^3 - 5n}{3} + C$$

with $C = 1$ for odd n and $C = 2$ for even n.

m	n	{n/2}	C
even	even	—	2
odd	odd	—	1
even	odd	even	3/2
even	odd	odd	1/2
odd	even	even	0
odd	even	odd	1

{n/2} indicates the greatest integer contained in the { }.

Figure 43   Table for the value of C

Figure 44 (from Hartmann) gives the maximum-length rook tours for values of m and n through 12. Neither Peters nor Hartmann attempted the much more difficult task of finding a formula for the number of distinct tours on a given board.

n \ m	2	3	4	5	6	7	8	9	10	11	12
1	2	4	8	12	18	24	32	40	50	60	72
2	4	8	16	24	36	48	64	80	100	120	144
3		14	26	38	56	74	98	122	152	182	218
4			38	54	78	102	134	166	206	246	290
5				76	104	136	174	216	264	316	374
6					136	174	220	270	328	390	460
7						218	270	330	396	470	550
8							330	396	474	556	650
9								472	558	652	756
10									652	756	872
11										870	996
12											1,134

Figure 44   Maximum-length rook tours for m × n boards from 1 × 2 through 12 × 12

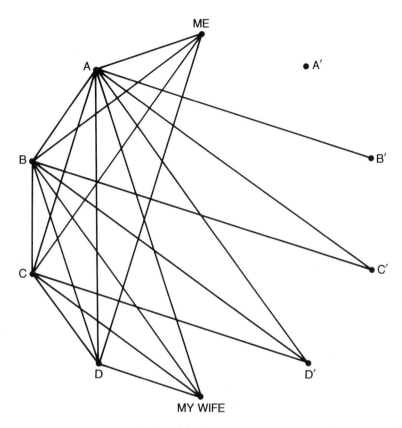

Figure 45   Answer to the handshaking problem

2. Among the five married couples no one shook more than eight hands. Therefore if nine people each shake a different number of hands, the numbers must be 0, 1, 2, 3, 4, 5, 6, 7 and 8. The person who shook eight hands has to be married to whoever shook no hands (otherwise he could have shaken only seven hands). Similarly, the person who shook seven hands must be married to the person who shook only one hand (the hand of the person who shook hands only with the person who shook eight hands). The person who shook six must be married to the person who shook two, and the person who shook five must be married to the person who shook three. The only person left, who shook hands with four, is my wife.

The above reasoning, which makes use of the familiar "pigeonhole principle," can be clarified by diagramming the problem [*see Figure 45*]. Every graph that lacks loops and multiple edges must contain at least two points that have the

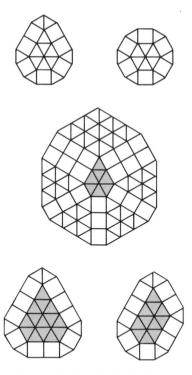

Figure 46   Eleven-sided and 12-sided convex polygons
and three other polygons of 11 sides

same number of lines attached to them. In this case the graph has only two such points, those representing me and my wife.

3. An 11-sided convex polygon can be formed with unit-sided squares and equilateral triangles, as shown in Figure 46, *[top left]*. The angles possible for a convex polygon formed with the pieces are 60, 90, 120 and 150 degrees. For a polygon with the maximum number of sides, all angles must be 150 degrees. The number of sides will then be 12. Figure 40 *[top right]* shows the smallest example.

Several readers "proved" that an 11-sided polygon could not be formed with squares and equilateral triangles of unit sides. The flaw, of course, was failing to realize that a side could be more than one unit long.

Wade Philpott pointed out that *any* convex pentagon formed with unit equilateral triangles can be used as the core of an 11-sided polygon. Simply place unit squares next to each triangle and complete the perimeter with six triangles. The solution I gave leads to an infinite family of 11-sided polygons, shown in the middle of Figure 46. At the bottom are two other examples with different inner pentagons. The problem derives from one posed by Joseph Malkewitch in *Mathematics Magazine* and answered by Michael Goldberg in the May 1969 issue, page 158.

4. Only the ninth statement is true.

David L. Silverman contributed the problem to the *Journal of Recreational Mathematics,* January, 1969, page 29, presenting it in the form of 1,969 statements. Underwood Dudley answered it in the October issue, page 231, as follows: "At most one of the statements can be true because any two contradict each other. All the statements cannot be false, because this implies that the list contains exactly zero false statements. Thus exactly one statement can be true. Thus exactly $n - 1$ are false, and the $(n - 1)$st (the 1,968th) statement is true."

Alan Brown pointed out that if the word "exactly" is removed from each of the 10 statements in the logic problem, there is a different and unique solution: The first five statements are true; the last five, false.

The problem obviously generalizes to as many statements as you care to add. What happens if you *decrease* the number to just one?

1. Exactly one statement on this list is false.

Norman Pos wrote to point out that the problem then reduces to the traditional liar paradox: "This sentence is false." To circumvent the paradox, Pos added a zero statement at the beginning:

0. Exactly none of the statements on this list is false.

Pos was surprised to discover that this shifts the one true statement from position $n - 1$ to position n. That adding such a statement at the top of, say, 1,000 numbered statements would shift the unique true sentence from next-to-last to last he found an amusing case of syntactical "action at a distance."

5. A solution to the farm problem, enclosing 128 square units, is shown in Figure 47.

I learned later that this problem had been proposed by R. J. French in *The Fairy Chess Review,* Vol. 4, 1939, page 43. French said the area was more than 120. I have not been able to determine if the problem was answered in subsequent issues.

After I published Knuth's 128 solution, Yoichi Kotani sent a proof, along with 1,440 solutions, that 128 is the maximum. Robert Reid Dalmau of Lima, Peru, sent the same set of solutions. In 1978 Takakazu Shimauchi published in Japanese a proof that 128 is the maximum (*Sugaku Seminar,* March, 1978, pages 11 – 16).

For references on pentomino farm problems in the *Journal of Recreational Mathematics,* see the issues for January, 1968, pages 55 – 61; October, 1968, pages 234 – 235; July, 1969, pages 187 – 188; and Vol. 17, No. 1, 1984 – 1985,

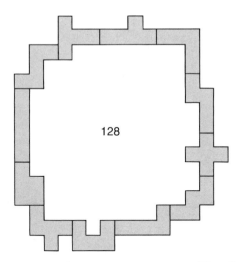

Figure 47   The largest pentomino "farm"

pages 75 – 77. If the 12 pieces are allowed to touch only at corners and all edges are required to be horizontal and vertical, a farm of 160 square units is the largest known. If the pieces are allowed any orientation and corner touching, the area can be raised to slightly more than 161.

6. A square table can always be placed somewhere on a wavy floor with all four legs touching the floor. To prove this, put the table anywhere. Assume that only three legs, *A*, *B*, *C*, are on the floor and *D* is off *[see Figure 48]*. It is always possible for three legs to touch the floor because three points, anywhere in space, mark the corners of a triangle. Rotate the table 90 degrees around its center, keeping legs *A* and *B* always on the floor. This brings the table to a position where *C* is now the only leg that does not touch the floor.

During the rotation *D* has moved to the floor and *C* has left it. But *D* must have touched the floor before *C* left, otherwise there would be a position at which only *A* and *B* would touch the floor, and we know that it is always possible for three legs to touch. At some point in the rotation, therefore, all four

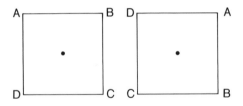

Figure 48   The wobbly-table proof

legs must have been in contact with the floor. A similar argument can be applied to wobbly rectangular tables by giving them 180-degree rotations.

Many readers called attention to two tacit assumptions that are necessary to make this proof valid:

1.   The table, like all normal tables, has four legs of equal length, their lower ends at the corners of a square.

2.   The legs of the table are sufficiently long and the unevenness of the floor is sufficiently mild, so while the table is rotated, there is never a moment at which three legs cannot be made to touch the floor.

The theorem is actually useful. Suppose you have a circular table with four legs that wobbles a bit when you move it to a porch. If you don't mind the table's surface being on a slight slant, you don't have to search for something to slip under a leg: Just rotate the table to a stable position. If you have to stand on a four-legged stool or chair to replace a light bulb and the floor is uneven, you can always rotate the stool or chair to make it steady.

7. To put a chicken-wire pattern of creases into a small sheet of paper, first roll the sheet into a tube about half an inch in diameter. With the thumb and forefinger of your left hand, pinch one end of the tube flat. Keeping pressure on the pinch with your left hand, your right thumb and forefinger, pinch the tube flat at a spot as close as possible to the first pinch, making the pinch at right angles to the first one. Press firmly with both hands, at the same time pushing the two pinches tightly against each other to make the creases as sharp as possible. Now the right hand retains its pinch while the left hand makes a third pinch adjacent to and perpendicular to the second one. Continue in this way, alternating hands as you move along the tube, until the entire tube has been pinched. (Children often do this with soda straws to make "chains.") Unroll the paper. You will find it hexagonally tessellated in a manner that is most puzzling to the uninitiated.

John H. Coker wrote to say that when he was a child in Yugoslavia in the early 1930's, his schoolteacher rolled and pinched notes to other teachers in this manner. Because it is extremely difficult to unroll such a tube and then re-roll it exactly as before, the tube provided security from the eyes of children asked to transmit the notes.

8. Place the pieces as shown in Figure 49, and make the following moves:

White	Black
1	B-Q4 (check)

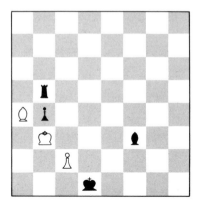

Figure 49   Retrograde chess

White	Black
2 P-B4	P takes P
	en passant
	(double check)
3 K takes P (check)	

Removing the White king will now leave the position given with the statement of the problem.

In addition to books of philosophical essays and logic problems, Raymond Smullyan published two collections of his marvelous chess problems: *The Chess Mysteries of Sherlock Holmes* (Knopf, 1979) and *The Chess Mysteries of the Arabian Knights* (Knopf, 1981).

9. The key to simplifying the three polypower equations is the basic law

$$(a^b)^c = a^{(b \times c)}$$

Applying this to the first equation gives

$$x^{(x^{x^x})} = (x^x)^x$$
$$x^{(x^{x^x})} = x^{(x^2)}$$

The two bottom *x*'s are equal; therefore their parenthetical exponents are equal. Cancel the bottom *x*'s and repeat the procedure:

$$x^{(x^x)} = x^2$$

The bottom *x*'s again drop out, leaving $x^x = 2$, which gives *x* the value 1.55961+.

The same procedure simplifies the second equation (down-4 equals up-4) to $x^x = 3$, and $x = 1.82545+$. Each succeeding equation increases the value of x^x by 1. The third equation reduces to $x^x = 4$, or $x = 2$.

The general procedure is to replace the up ladder by a number one less than the number of its x's and remove two x's from the down ladder. (Example: Down-5 = up-5 reduces to down-3 = 4.)

Correspondence about polypowers was unusually heavy, and readers raised many interesting questions. Several readers pointed out that parentheses could be placed on a ladder in a variety of ways. In fact, the number of ways is given by the sequence known as the Catalan numbers. However, not all ways of parenthesizing give distinct values for the ladder. Determining the number of such values is a difficult problem, and I do not know the solution.

Many readers called attention to unusual, little-known theorems about infinite ladders of exponents. Consider, for example, a ladder of x's that grows steadily upward to infinity. I would have thought that if x is greater than 1 the ladder's value (working from top down) would diverge as the ladder grows. This is not true. If x is an integer, the value diverges only if x exceeds $e^{1/e} = 1.4446. \ldots$. If x is a real number, it converges only if it is equal to or greater than $e^{-e} = 0.0659 \ldots$ and equal to or less than $e^{1/e}$. I found this amazing.

A delightful paradox is related to the above theorem. Assume that an infinite ladder of x's has a value of 2. What is the value of x? Because all the x's above the bottom x form an infinite chain, we can assume that the value of this chain is also 2. Substituting 2 for this chain gives the equation $x^2 = 2$, for which $x = \sqrt{2}$.

All well and good. Now apply the same dodge to an infinite ladder of x's that equals 4. This leads to $x^4 = 4$, so again $x = \sqrt{2}$. How can an infinite ladder converge to two different numbers? Actually, an infinite ladder of square roots of 2 cannot converge to 4, and in this case the dodge is not applicable. To show this exactly is complicated. You will find it explained in "A Matter of Definition," by M. C. Mitchelmore in *American Mathematical Monthly*, Vol. 81, 1974, pages 643–647.

For general discussions of infinite ladders see "Infinite Exponentials," by D. F. Barrow in *American Mathematical Monthly*, Vol. 43, 1936, pages 150–160; "Exponentials Reiterated," by R. A. Knoebel, *ibid.*, Vol. 88, 1981, pages 235–252; and "Infinite Exponentials," by P. J. Rippon in *Mathematical Gazette*, Vol. 67, 1983, pages 189–196. Knoebel gives a long bibliography of earlier references.

Several readers sent references relevant to Grosse's labors, but unfortunately they were all in German or French. I still know of no good references in English to the sort of problems Grosse has been investigating.

Some comments on big numbers may be of interest. I mentioned that the largest number that can be written in conventional notation with no symbols other than three digits is 9^{9^9}. In the next-to-last chapter of *Ulysses,* Joyce reveals that Leopold Bloom was once fascinated by this number, and a paragraph is devoted to describing how big it is.

Skewes is pronounced Skew-ease. The large number that bears his name was based on the assumption that the Riemann hypothesis is true. What if it isn't? In 1955 Skewes published a proof that the number would then be the much larger

$$10^{10^{10^{10^3}}}$$

For an entertaining account of all this see "Skewered!" by Isaac Asimov in *Fantasy and Science Fiction,* November, 1974. Skewes made his calculations at the request of J. E. Littlewood, who tells about it in the chapter titled "Large Numbers" in *A Mathematician's Miscellany* (Methuen, 1953).

Even Skewes's second number is very tiny and no longer the largest ever involved in a legitimate proof. The record is now held by Ronald L. Graham, of Bell Laboratories. Graham's number arose in connection with a problem in a branch of graph theory called Ramsey theory. (See my *Scientific American* column for November 1977.) The number can be expressed compactly only in a special notation devised by Donald E. Knuth for handling numbers of such unimaginable magnitude.

CHAPTER SEVEN

Napier's Bones

In his celebrated *Budget of Paradoxes* Augustus De Morgan defines a "grapho-math" as a person ignorant of mathematics who tries to describe a mathematician. As an example, he quotes from the second chapter of Sir Walter Scott's novel *The Fortunes of Nigel,* in which David Ramsay, a whimsical clockmaker and amateur mathematician, swears "by the bones of the immortal Napier!"

It is hard to tell from the passage whether Scott actually was uninformed or whether he merely intended Ramsay to make an ignorant or a joking remark. In any case, "Napier's bones" have nothing to do with the skeletal remains of Baron John Napier (1550–1617), the Scottish mathematician who discovered logarithms and who was the first important mathematician of Britain. The phrase refers to a set of numbered rods that Napier invented for doing multiplication. We shall discuss his method later, but first some remarks about Napier himself.

His father, Sir Archibald Napier, master of the Scottish mint, was just 16 when John was born. And John was a mere 13 when he entered the University of St. Andrews. He left the university without getting a degree, took over the family castle and estates at Merchiston (now part of Edinburgh), married and had one son and one daughter, was widowed, remarried and continued the symmetry with five sons and five daughters. The Protestant Reformation in Scotland had started at about the time John was born, and while a youth at St. Andrews he became a passionate Calvinist with a compulsion to explicate biblical prophecy. In 1593 he published what he always considered his master-piece (much more important than logarithms), the full title of which was "A Plane Discovery of the whole Revelation of Saint Iohn: set downe in two treatises: The one searching and proving the true interpretation thereof: The other applying the same paraphrastically and historically to the text. Set foorth by John Napier L. of Marchistoun younger. Whereunto are annexed certaine Oracles of Sibylla, agreeing with the Revelation and other places of Scripture.

Edinburgh, printed by Robert Walde-grave, printer to the King's Majestie, 1593. Cum privilegio Regali."

It was the first major Scottish work on the Bible and one of the most thorough attempts ever made before or since to explore the symbolism of the Apocalypse. It is ironic that today, when many college students seem more interested in the Second Coming than in current politics, there is no available reprint of Napier's treatise. It was enormously influential in its day, with 21 English editions and numerous European translations.

Perhaps the main reason the book is out of print is that Napier made a slight miscalculation about the end of the world. He had been strongly influenced by the religious speculations of Michael Stifel, a German algebraist who proved that Pope Leo X was the Antichrist by rearranging the Roman numerals in LEO DECIMVS to make DCLXVI, or 666, the notorious "mark of the Beast." Where did Stifel get the x? from Leo X and because LEO DECIMVS has 10 letters. What happened to the M? He left that out because it stood for *mysterium*. Stifel predicted that the world would end on October 3, 1533. Napier perceived that this was a mistake. He decided that it was the Pope of 1593 who was the actual Antichrist. God had ordained that exactly 6,000 years would elapse between the earth's creation and its destruction. Since there was some uncertainty about the exact date of creation, Napier set the end of the world as being between 1688 and 1700.

Napier begins his book by apologizing for having written it in a language so base as English, and he concludes it by appealing to the pope as follows:

"In summar conclusion, if thou o *Rome* aledges thyselfe reformed, and to beleeue true Christianisme, then beleeue Saint *John* the Disciple, whome Christ loued, publikely here in this Reuelation proclaiming thy wracke, but if thou remain Ethnick in thy priuate thoghts, beleeuing the old Oracles of the *Sibyls* reuerently keeped somtime in thy *Capitol:* then doth here this *Sibyll* proclame also thy wracke. Repent therefore alwayes, in this thy latter breath, as thou louest thine Eternall salvation. *Amen.*"

"Strange," comments De Morgan in his *Budget*, "that Napier should not have seen that this appeal could not succeed, unless the prophecies of the Apocalypse were no true prophecies at all."

After clearing up the apocalyptic mysteries, Napier turned his ingenuity toward ways of defending Scotland against a threatened invasion by Catholic Spain. His 1596 document was titled *Secrett Inventionis, proffitabill and necessary in theis dayes for defense of this Iland, and withstanding of strangers, enemies of God's truth and religion*. It describes three inventions: mirrors for setting fire to enemy ships (shades of Archimedes!), a machine gun and a metal chariot (that is, a tank) housing soldiers who could fire through holes in the sides.

Napier's next book, the Latin title of which begins *Mirifici Logarithmorum Canonis Descriptio . . . (A Description of the Marvelous Rule of Logarithms . . .)*, appeared in 1614. This was the book in which Napier explained logarithms, called them logarithms (a term he coined) and gave the world its first log table. It has often been pointed out that if exponents had then been in common use, logarithms would have immediately been recognized as a great toil saver, but Napier conceived of them without reference to exponents at all. This is not the place to explain how he arrived at logs the hard way by considering the relation of an arithmetic series to a geometric series. The London geometer Henry Briggs quickly realized that 10 was the most convenient base for logarithmic calculations in the decimal system, and Napier at once agreed. It is said that when the two men first met at Merchiston Castle (where Briggs remained for a month), they admired each other for 15 minutes before either spoke a word.

Navigators and astronomers, notably Johannes Kepler, found the base-10 logs (or common logarithms as they are now called) invaluable, and years of drudgery were devoted by Briggs and others to preparing better and better log tables. (Today it is faster to compute a log all over again on a pocket electronic calculator—it takes less than a second—than to look it up in a book!) In Napier's posthumous work *Mirifici Logarithmorum Canonis Constructio . . .* (1619) he explained how he calculated his original logs. In doing so he made systematic use for the first time in history of a decimal point, placing it above the baseline and using it exactly as it is used today in England.

Two of the most amusing of many anecdotes about Napier are recounted by Howard W. Eves in his delightful *In Mathematical Circles*. Because a neighbor's pigeons were flying onto Napier's estate and eating grain, Napier told his neighbor that he would impound the birds as payment. The neighbor replied that Napier was welcome to any pigeon he could catch alive. Napier scattered brandy-soaked peas over his grounds and the pigeons were soon staggering about in such a stupor that he had no trouble collecting all of them in a sack.

It was a time when almost everyone in Scotland (including Napier) believed in astrology and black magic. One day Napier called his servants together and told them that his black rooster had the occult power to tell him which servant had been stealing from the estate. One at a time each servant was asked to enter a dimly lighted room and stroke the bird's back. As Napier had anticipated, only the guilty person, fearing exposure, would not do as asked. Napier had covered the rooster's black feathers with soot, and so only the guilty servant emerged with clean hands.

The age was also one of intense interest in calculating. The average person did arithmetic on his fingers, but more skillful mathematicians took great

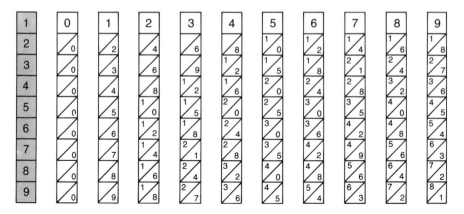

Figure 50   Rabdology, or "Napier's bones"

delight in completing tedious computations. Napier's hobby was to find ways to simplify such work. Logarithms were, of course, his best invention, but in 1617 (the year he died) he brought out a little book called *Rabdologia* that explained three other methods of calculating. The book's title was his name for the first method, one that soon became known as "Napier's bones" because it used rods that often were made of animal bone.

The reader is urged to make a set of Napier's bones by labeling 11 strips of heavy cardboard (or Popsicle sticks, tongue depressors or any other available wooden strips) as shown in Figure 50. The index rod is not essential, but it makes it easier to locate desired rows. Each of the rods has a digit at the top. Below the digit, from the top down, are the products when that digit is multiplied successively by numbers 1 through 9. The set of bones obviously is nothing more than a multiplication table cut into strips so that it can be manipulated manually, with a zero strip added to serve as a placeholder.

The procedure is ridiculously simple. Suppose you wish to multiply 4,896 by 7. Rods topped with 4, 8, 9 and 6 are placed side by side with the index rod on the left *[see Figure 51]*. Only row 7 (the multiplier) is considered. Write down 2, the last digit of the row, as the final digit of the product. The product's next digit (working to the left on both rods and paper) is obtained by adding the next pair of digits (the diagonally adjacent digits inside the little parallelogram) of the row. They are 4 + 3, so put down 7 as the second digit from the end of your product. The sum of the next pair (6 + 6 = 12) is more than 9, therefore write 2 as the third digit of the product and carry 1. The next pair, 5 and 8, add to 13, but you are carrying 1, so the sum is 14. Put down 4 and again carry 1. The last digit of the row is 2. Two plus 1 is 3, so 3 is the final digit (on the left) of your product.

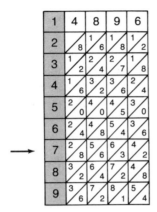

Figure 51 $4{,}896 \times 7 = 34{,}272$

You have now obtained the correct answer, 34,272, by using only simple addition. Of course, if you know your multiplication table through the 9's, you can do it just as easily without the rods. In Napier's day, however, the ordinary person's ability to calculate was feeble, so the rods became an instant success throughout Great Britain and continental Europe.

To multiply 4,896 by a larger number, say 327, it is necessary to obtain three partial products and add them in the usual way. In other words, write down 34,272 (the product of 4,896 and 7); then put below it the products obtained from rows 2 and 3, jogging them to the left in the standard manner,

$$34272$$
$$9792$$
$$\underline{14688}$$

then add to obtain the final product.

The rods are of little use unless you have more than one set because a multiplicand may contain duplicate digits. Napier's rods had square cross sections, each face of a rod corresponding to one of the strips in our cardboard set. He arranged the four columns so that the top digits on opposite sides of each rod added to 9. The following are the quadruplets of Napier's set of 10 bones:

0, 1, 9, 8	1, 3, 8, 6
0, 2, 9, 7	1, 4, 8, 5
0, 3, 9, 6	2, 3, 7, 6
0, 4, 9, 5	2, 4, 7, 5
1, 2, 8, 7	3, 4, 6, 5

It is clear that such a set of 10 rods can handle all multiplicands of 10 digits or fewer that are possible to form with the rods, but many multiplicands cannot be formed, so it was advisable to own more than one set. As a little puzzle in combinatorics, can the reader determine the largest multiplicand one set of Napier's bones will form such that all smaller multiplicands can also be formed by the set? As a second exercise, find the corresponding largest multiplicand for two sets of Napier's bones.

Napier's rods can be used for division too, but the process is more trouble than it is worth. In short division you must select rods that form the desired dividend on the row for the digit divisor and read off the quotient from the top.

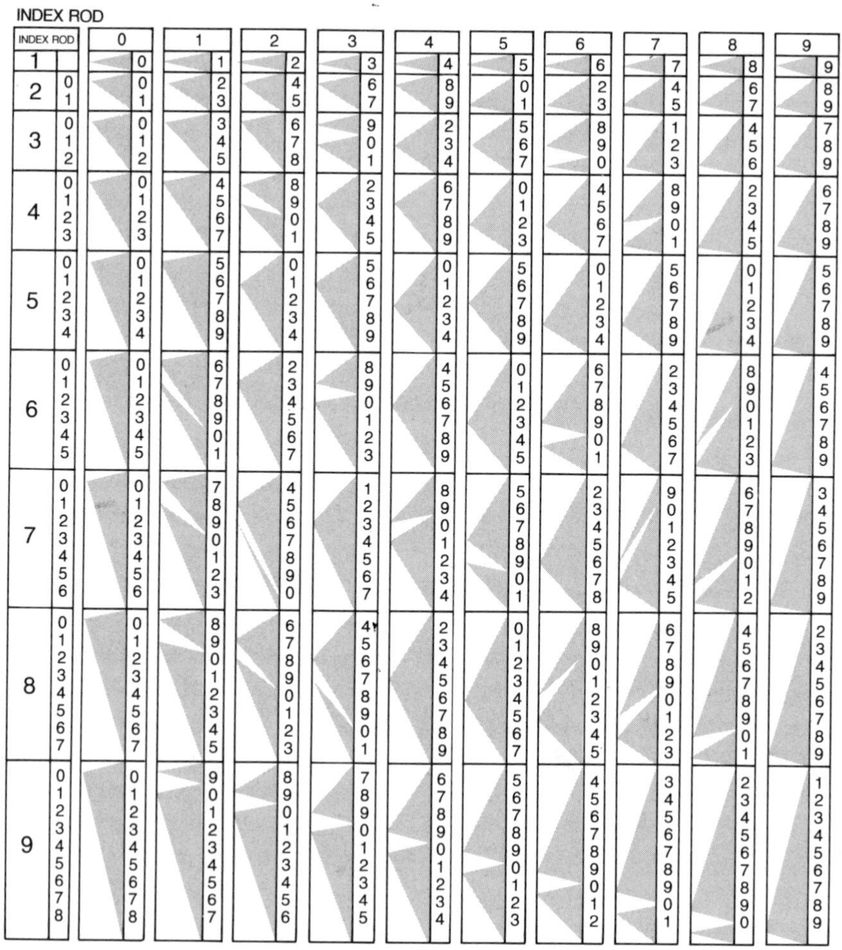

INDEX ROD

INDEX ROD	0	1	2	3	4	5	6	7	8	9

Figure 52   Henri Genaille's calculating rods

INDEX ROD	6	7	3

8	0	8	6	4
	1	9	7	5
	2	0	8	6
	3	1	9	7
	4	2	0	8
	5	3'	1	9
	6	4	2	0
	7	5	3	1

Figure 53   673 × 8 = 5,384

If the dividend cannot be formed, form the largest number you can that is less than the dividend and subtract that number from the dividend to obtain the remainder. In long division the rods can be used for determining the successive products of the divisor and each digit in the quotient.

The charm of Napier's rods lies in their simplicity. If we are willing to complicate them a bit, however, we can eliminate the bother of having to carry 1's in our head. The cleverest way of doing this was invented about 1890 by Henri Genaille, a French civil engineer. The picture of these rods is almost self-explanatory *[see Figure 52]*. They work exactly like Napier's except that the product is read directly from right to left. Start with the digit at the top right of the desired row. The next digit is the one to which the shaded triangle (at the left of the previous digit) points. From now on move from each digit into the shaded triangle directly at its left and go to the digit to which it points. For example, to multiply 673 by 8, start with 4 at the top right *[see Figure 53]*, and see how easily you can move to the left through the chain of triangles to obtain the product 5,384.

Both Napier's bones and Genaille's rods are marvelous teaching devices because it is not hard to see why they work, and when you do, you obtain valuable insight into the multiplication procedure. If you have difficulty understanding why Genaille's rods operate, you can find it explained in the article by B. R. Jones (see the bibliography), from which our illustrations were taken.

The second calculating method in *Rabdologia* had to do with arranging metal plates inside a box. It is too complicated and impractical to explain here. But Napier's third method, which he regarded as being primarily an amusement, requires only a chessboard and a supply of counters. By moving the counters as you would rooks or bishops, you can do addition, subtraction, multiplication, division and square roots all in the binary system.

ANSWERS

A single set of Napier's original 10 rods will form every multiplicand of 11,110 or less. Two sets will form every multiplicand of 111,111,110 or less. For n sets of bones the number is $4n$ 1's followed by a 0.

ADDENDUM

Napier had no notion of a "base" for his logarithms. The matter is complicated, but, as Carl Boyer explains in his *History of Mathematics,* if you divide all Napier's numbers and logarithms by 10^7, you have a system practically the same as one based on $1/e$. Natural logs based on e later came to be known as Napierian logarithms, even though Napier never had such a system. Boyer does a good job clarifying the confusing details.

Napier's bones were based on an ancient way of multiplying that came to be called the Gelosia system, because its lattice lines looked like the gratings on Italian windows. There is a good account of this, together with a survey of the curious mechanical devices (some with rotating cylindrical rods) that came after Napier's bones, in the paper by M. R. Williams cited in the bibliography.

I had assumed that David Ramsay, mentioned in the first paragraph of the chapter, was invented by Scott. Not so. He actually lived and made and sold clocks and watches for a living. He served James I as an astrologer, as did his son William. In 1652 William published a book on astrology with a curious dedication to his father that reads in part: "It's true your carelessness in laying up while the sun shone for the tempests of a stormy day hath given occasion to some inferior-spirited people not to value you according to what you are by nature and in yourself, for such look not to a man longer than he is in prosperity. . ."

William Lilly, a famous British astrologer of the time, wrote an autobiography in which he gives a hilarious account of how he, David Ramsay and others tried to locate a treasure reportedly buried in the cloisters of Westminster Abbey. It was late at night, and a great wind developed that prevented their dowsing rods from turning. Lilly writes that he "dismissed the demons," but that the real cause of their failure was that they were surrounded by more than 30 people who kept laughing and deriding them. Lilly's autobiography is also the source of the anecdote I gave about the first meeting of Napier and Briggs.

Where today are Napier's body bones? Nobody seems to know. As Williams discloses, there are reports of his having been buried in at least two different spots in Edinburgh.

BIBLIOGRAPHY

The Art of Numbering by Speaking-Rods: Vulgarly Termed Napier's Bones. W. Leybourn. London, 1667.

"John Napier" and "Logarithms." J. W. L. Glaisher in *The Encyclopedia Britannica*, 11th edition, 1911.

Napier Tercentenary Memorial Volume. Edited by Cargill Gilston Knott. Longmans, 1915.

"Lord Napier — First Scottish Expositor of the Revelation." Leroy Edwin Froom in *The Prophetic Faith of Our Fathers*, Vol. 2. Review and Herald, 1948. Froom is a Seventh-Day Adventist. His section on Napier is the best account of Napier's eschatology that I know.

"Genaille's Rods: An Ingenious Improvement on Napier's." B. R. Jones in *The Mathematical Gazette*, Vol. 48, 1964, pages 17–22.

"John Napier and the History of Logarithms." N. T. Gridgeman in *Scripta Mathematica*, Vol. 29, 1969, pages 49–65.

"From Napier to Lucas: The Use of Napier's Bones in Calculating Instruments." M. R. Williams in *Annals of the History of Computing*, Vol. 5, 1983, pages 279–296; see also comments in Vol. 6, 1984, pages 403–404.

"Napier's Bones." Michael R. Williams in *A History of Computing Technology.* Prentice-Hall, 1985.

CHAPTER EIGHT

Napier's Abacus

"Napier's bones" (the topic of the previous chapter) are the calculating rods that were invented by John Napier, the 16th-century Scottish mathematician who discovered logarithms. In *Rabdologia,* the book in which Napier first explained his "bones," he also described a curious method of calculating by moving counters across a chessboard. This method, which seems to have been completely forgotten, deserves to be remembered for several reasons: It is not only a pleasant recreation but also a valuable teaching device and of considerable historic interest. It is the world's first binary computer, and it came almost 100 years before Leibniz explained how to calculate with binary numbers! Although Napier did not express numbers explicitly in binary notation, we shall see how his counting board is equivalent to doing so.

The use of checkered boards and cloths for calculating was widespread in Europe during the Middle Ages and the Renaissance. English words such as "exchequer," "check" and "counter" derive from these boards; even "bank" comes from the German word for counting board, *Rechenbank*. The algorithms for calculating on these boards, however, were clumsy. By adopting a binary system and basing his algorithms on old methods of multiplying by "doubling," Napier created a remarkably efficient counting board unlike any that had been in use before.

Napier's counting board is a chess-board of arbitrary size, with columns and rows labeled by the doubling series 1, 2, 4, 8, 16, 32,. . . . These numbers are, of course, successive powers of 2. Before explaining Napier's methods for multiplying, dividing and extracting square roots, let us see how his board can be used for addition and subtraction. Suppose we want to add 89 + 41 + 52 + 14. Each number is expressed by placing counters on a row of the board *[see Figure 54a]*. A counter has the value of its column. (Ignore the row numbers on the right margin.) Thus the fourth row shows 89 as the sum of 64 + 16 + 8 + 1. If you think of each counter as 1 and the empty cells as 0, then 89 is

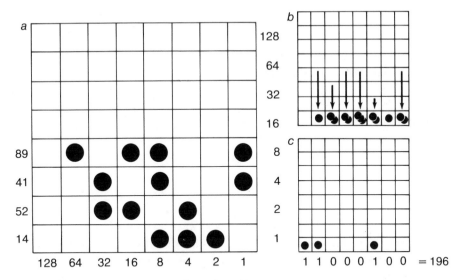

Figure 54 Binary addition: 89 + 41 + 52 + 14

represented in binary notation as 1011001, and similar notations can be made for the other three numbers. The counters can be positioned rapidly because any positive integer is uniquely represented as a sum of the powers of 2. Start at the left and put a counter on the largest power less than the number to be represented; then move right and place a counter on the next larger power that, when added to the previous power, will not exceed the desired number. Continue in this way until the unique binary representation is obtained.

To add the four numbers, first move all the counters down like rooks in chess to the bottom row [see Figure 54b]. Adding the values of all these counters will give the correct sum, but we want to express the sum in binary notation. To do this, "clear" the row of multiple counters on a cell by the following procedure. Start at the right, taking each cell in turn. Remove every pair of counters on a cell and replace them with a single counter on the adjacent cell to the left. We shall call this "halving up." Clearly it will not affect the sum of the counters's values because every pair of counters of value n is replaced by one counter of value $2n$. The final result after clearing is the binary number 11000100, or 196 in decimal notation [see Figure 54c].

Subtraction is almost as simple. Suppose you want to take 83 from 108. Represent the larger number on the second row and the smaller on the bottom row, as shown in Figure 55a. You can now do subtraction in the usual manner, starting at the right and borrowing as you go, but I prefer to alter the entire second row (preserving the total value of its counters) until each counter on the bottom row has one or two counters above it, and no empty cell on the bottom

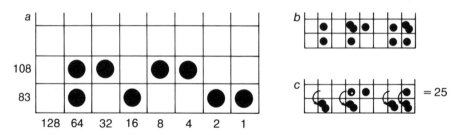

Figure 55   Binary subtraction: 108 − 83

row has more than one counter above it. This is done by "doubling down" on the second row — removing a counter and replacing it with two counters on the next cell to the right. How the top row looks after it is transformed to meet the two specified conditions is shown in *b*. The next step is to "king" (as in checkers) each counter on the bottom row by moving a counter on top of it from the cell directly above. After this is done, the second row shows in binary notation the difference between the two numbers. In this case it is 11001, or 25, as shown in Figure 55*c*.

A different subtraction method is to "complement" the smaller number and add. A number is complemented by placing a counter on each of its empty cells and removing all counters originally there. In other words, each 0 is changed to 1, and each 1 to 0. (If the subtrahend has fewer digits than the minuend, before complementing you must add zeros to the left of the subtrahend until it is the same length as the minuend.) Add the two numbers, clear the row by halving up and transfer the counter at the extreme left to the extreme right. Clear again if necessary. To use the preceding example, we change 1010011 to its complement 101100. Adding and clearing produces 10011000. Shifting the counter from left to right gives 11001, or 25, the correct difference.

Multiplication is delightfully easy. Napier explains it with the example 19 × 13 = 247. One number, say 19, is indicated below the board by marking the proper columns; the other number, 13, by marking the proper rows. A counter goes on each cell at the intersection of a marked row and column *[see Figure 56a]*. Every counter not on the column at the extreme right is moved diagonally up and right (like a chess bishop) until it is on the rightmost column. The result is shown in *b*. The sum of the values of these counters (as indicated on the right margin) is 247, the desired product, but we wish to express it in binary notation. That is quickly done by halving up until the column is cleared. The final result is 11110111, or 247, as shown in *c*.

It is easy to see why it works. Counters on the first row keep their values when moved to the right, counters on the second row double in value, counters on the

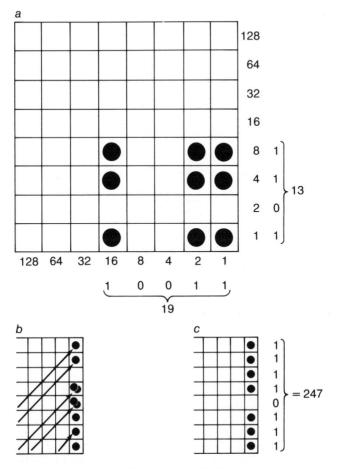

Figure 56 Binary multiplication: 19 × 13

third row quadruple in value and so on. The procedure is equivalent to multiplying with logarithms to base 2. In our example, 19 is expressed as $2^4 + 2^1 + 2^0$, and 13 as $2^3 + 2^2 + 2^0$. Cross multiplying in the familiar manner (remembering the basic law of exponents: $x^n \times x^m = x^{n+m}$) yields $2^7 + 2^6 + 2 \cdot 2^4 + 2 \cdot 2^3 + 2^2 + 2^1 + 2^0$. This corresponds exactly to Napier's procedure. Indeed, moving the counters is equivalent to cross multiplying. We are, in effect, multiplying by adding exponents.

Napier was not the first to recognize that powers of 2 can be multiplied by adding their exponents. As early as 1500 it had been clearly explained with exponential notation by Nicolas Chuquet, a physician of Lyons, in the algebraic part of his *Triparty en la sciences des nombres*. It is Napier, however, who gets the credit for the first mechanical device operating with logs based on 2.

Napier next explains how to do long division on his abacus, using the example 250 ÷ 13. The procedure, as one would expect, is the reverse of

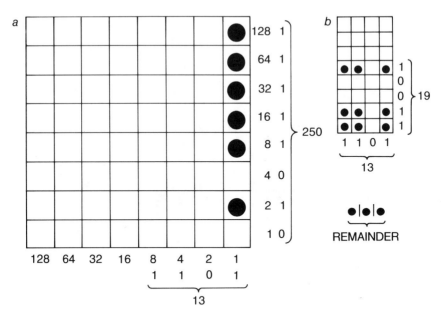

Figure 57 Binary division: 250 ÷ 13

multiplication. Complications arise that make it difficult to explain, although in practice one soon learns to do it quickly. The divisor, 13, is marked at the bottom of the board, and the dividend is indicated by counters on the column at the extreme right *[see Figure 57a]*. You must now move the dividend counters like chess bishops, down and left, to produce a pattern that has counters (one to a cell) only on marked columns, and each marked column must have its counters on the same rows. Only one such pattern can be formed, but to do so it is necessary at times to double down on the right column, that is, remove single counters, replacing each with a pair of counters on the next lower cell.

Start with the top counter and move it diagonally to the leftmost marked column. If you see that you cannot proceed to form the desired pattern, return the counter to its original cell, double down and try again. If the first attempt fails, the second will succeed in beginning the required pattern, although more doubling down may be necessary. Continue in this manner, doubling down whenever you see that you must, gradually filling in the pattern by extending it down and right until finally the unique pattern is constructed *[see Figure 57b]*. After the final counter at the bottom right corner of the pattern is in place, you will have three counters left over. They represent the remainder. The rows containing counters are marked on the right margin, symbolizing 10011, or 19, the correct quotient. The three extra counters give the fraction ³⁄₁₃.

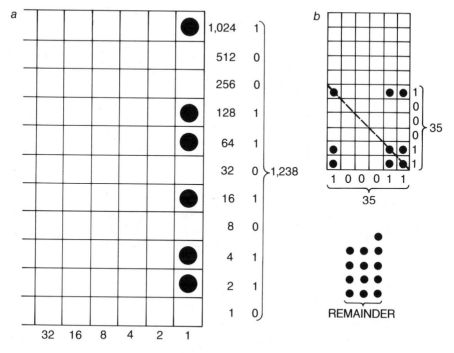

Figure 58   Binary extraction of square root: $\sqrt{1,238}$

A similar procedure is used to find integral square roots. If the root is not integral, the procedure gives the root of the largest square less than the original number. Counters left over then represent the difference between that number and the original. Napier demonstrates by finding the square root of the largest square less than 1,238. This requires a board extended higher than the standard chessboard. As in division, the number is represented by counters on the right-most column *[see Figure 58a]*. Since no divisor is marked on the bottom, how do we form a pattern? We must move counters diagonally down to produce a pattern with two properties: (1) Every column with counters must have its counters on the same rows, and (2) the pattern must have bilateral symmetry along the diagonal passing through the board's lower right corner. This ensures, of course, that multiplier and multiplicand are identical. As before, start with the top counter and see if you can move it to the diagonal of symmetry. If you can, that is the correct first move. If you cannot, double down and move one of the counters to the diagonal of symmetry. Continue in this fashion, doubling down when necessary, until the required symmetrical pattern is achieved. The result is $35 \times 35 = 1,225$, with 13 leftover counters that represent the difference between the square and 1,238.

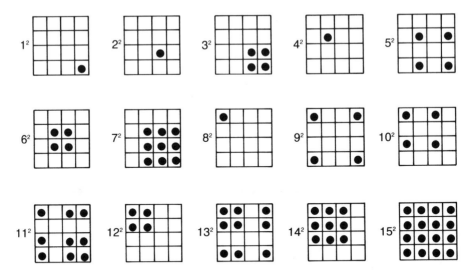

Figure 59   Patterns for squaring 1 through 15

The 15 patterns that generate all squares from 1 through 225 are shown in Figure 59. Studying them will familiarize you with the kind of pattern that must be formed for square roots. Note that in every pattern each row and column has a counter on the diagonal of symmetry.

Napier's device will operate with any base notation, but above base 2 it is necessary to work with multiple counters on single cells. As the base increases, the system becomes progressively more cumbersome and uninteresting, and more multiplying must be done in the head. For example, to multiply 77 by 77 in decimal notation each of the four cells at the lower right corner must hold $7 \times 7 = 49$ counters. After moving them to the right column you have 49 counters on the bottom cell, 98 on the next and 49 on the next. Then every set of 10 counters on a cell is replaced by a single counter immediately above it, resulting finally in counters on four cells that signify the product, 5,929.

The most interesting extension of Napier's board was suggested by Donald E. Knuth, the Stanford computer scientist. A checkered board can be used very efficiently for calculating in the "negabinary system." Because this remarkable notation is based on powers of -2, the rows and columns of the board are labeled with the series $+1, -2, +4, -8, +16, -32, \ldots$, in which alternate powers are negative. The main virtue of negabinary is that every positive and every negative integer can now be uniquely represented in binary notation without the use of signs. Examples are $13 = 11101$ $(16 - 8 + 4 + 1)$ and $-13 = 110111$ $(-32 + 16 + 4 - 2 + 1)$.

1	1	−1	1 1
2	1 1 0	−2	1 0
3	1 1 1	−3	1 1 0 1
4	1 0 0	−4	1 1 0 0
5	1 0 1	−5	1 1 1 1
6	1 1 0 1 0	−6	1 1 1 0
7	1 1 0 1 1	−7	1 0 0 1
8	1 1 0 0 0	−8	1 0 0 0
9	1 1 0 0 1	−9	1 0 1 1
10	1 1 1 1 0	−10	1 0 1 0
11	1 1 1 1 1	−11	1 1 0 1 0 1
12	1 1 1 0 0	−12	1 1 0 1 0 0
13	1 1 1 0 1	−13	1 1 0 1 1 1
14	1 0 0 1 0	−14	1 1 0 1 1 0
15	1 0 0 1 1	−15	1 1 0 0 0 1
16	1 0 0 0 0	−16	1 1 0 0 0 0
17	1 0 0 0 1	−17	1 1 0 0 1 1
18	1 0 1 1 0	−18	1 1 0 0 1 0
19	1 0 1 1 1	−19	1 1 1 1 0 1
20	1 0 1 0 0	−20	1 1 1 1 0 0

Figure 60   Negabinary notation of integers

The negabinary forms of positive and negative integers from 1 through 20 are shown in Figure 60. Note that every positive number has an odd number of negabinary digits and every negative number has an even number of negabinary digits. Every odd number, regardless of sign, ends in 1; every even number, regardless of sign, ends in 0. Many other basic theorems are easily discovered. For example, a negabinary number is divisible by 3 if and only if its number of 1's is a multiple of 3. Observe that every palindromic negabinary number on the list (a number that is the same in both directions) is a positive or a negative prime. Is this true in general? If not, what is the first exception?

I know of no better way to become acquainted with this extraordinary notation (so rich in recreational possibilities) than to calculate with it on Napier's board. Addition is handled exactly as before except that in clearing the sum the following two rules are observed:

1.   A pair of counters on one cell and a single counter on the next higher cell cancel one another. Remove all three.

2.   If any cells still have double counters, remove each pair and put single counters on each of the *two* next higher cells.

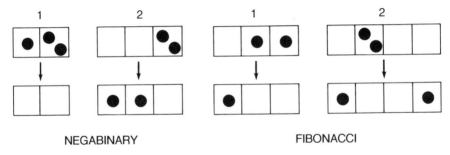

NEGABINARY   FIBONACCI

Figure 61   Clearing rules for negabinary and Fibonacci notations

The clearing procedure, thanks to the cancellation rule, is unusually rapid *[see Figure 61, left]*.

The fastest way to do subtraction is to change the sign of the subtrahend and add! Changing the sign is the same as multiplying by -1, or 11 in negabinary. Since multiplying by 11 is the same as adding a number to itself, with one replica shifted one cell to the left, we can reverse the sign of any negabinary number by the following simple algorithm: Add a new counter to every cell that is immediately to the left of a counter originally there, then clear the row as explained. For example, 11 (-1 in decimal notation) becomes 121, but the first two digits cancel (by rule 1), leaving 1, which is positive. Applying the algorithm again restores 11, or -1. When this algorithm is used on standard binary numbers, by the way, it is the same as multiplying by 3. (Do you see why?)

Any two negabinary numbers can be multiplied by using Napier's procedure and clearing the result according to negabinary rules. The product will have the correct sign when translated into decimal notation. Try multiplying -4 and -6. They are 1100 and 1110 in negabinary *[see Figure 62]*. After multiplying and clearing, you get 1101000, or $+24$. If you had multiplied -4 and $+6$, or $+4$ and -6, the result would have been 111000, or -24.

Division and square-root procedures are much trickier, although interested readers should be able to devise them. In square roots both positive and negative roots appear as solutions. Are there ways to use Napier's board efficiently for converting a signed binary number to negabinary, and vice versa? Yes; we can exploit two simple algorithms given by Knuth as the answer to Exercise 12 on page 177 of his *Semi-numerical Algorithms [see bibliography]*. Readers are encouraged to work them out before checking the answers section.

It is hard to believe, but the idea of negative-base notation (it applies to any radix) did not occur to anyone until the 1950's, when many people independently thought of it. In 1955, when Knuth was a high school senior, he wrote a short paper on it for a science talent search, but the first published account (at least in English) seems to be a short letter by Louis B. Wadel in *IRE Transactions on Electronic Computers* (Vol. EC-6, 1957, page 123). The term "negabi-

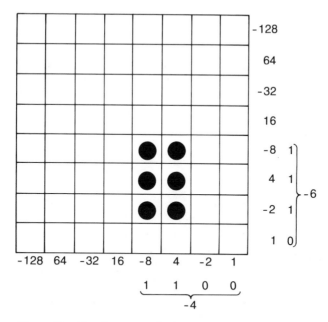

Figure 62   Negabinary multiplication: -4×-6

nary" was coined by Maurits P. de Regt, whose series of pioneering articles on negative radix arithmetic is listed in the bibliography.

Knuth also suggests the Fibonacci labeling, 1, 2, 3, 5, 8, 13, . . . , for Napier's board. It is difficult to multiply or divide with it, but addition and subtraction can be handled by representing each integer as the sum of the fewest possible Fibonacci numbers. Start by putting a counter on the column with the highest value less than the number to be represented; then work downward until the desired sum is obtained. (This method of representing a number uniquely in Fibonacci notation is known as Zeckendorf's theorem.) For example, 19 is uniquely indicated by 101001, or $13 + 5 + 1$. The adding procedure is the same as Napier's except that a row is cleared by the following two rules:

1.   If single counters are on adjacent cells of the board, remove them and put one counter on the next higher cell.

2.   For every pair of counters on the same cell, remove them and put one counter on the next higher cell and one on the *second* lower cell.

For example, two counters on cell 13 are replaced by one on cell 21 and one on cell 5 *[see Figure 61, right]*.

If you imagine the row extended two more cells to the right, with values of 1 and 0 (or, alternatively, that the columns are labeled 0, 1, 1, 2, 3, 5, . . .), then the above two rules suffice. Otherwise there are two exceptions. A pair of counters on 2 is replaced by one on 3 and one on 1, and a pair on 1 is replaced by one on 2.

To subtract, I know of no better way than the "kinging" procedure explained for binary subtraction. You must, of course, first change the minuend to the required pattern by applying the two clearing rules in reverse. There may be a better method. Indeed, there may be all kinds of clever algorithms for calculating on Napier's board, in various notations, that no one has yet discovered.

ANSWERS

To change a signed binary number to negabinary

1.   Express the number in binary on row 2.

2.   If the number is positive, move all counters that have negative values (in negabinary) down like rooks to the first row. (On a standard chessboard this means moving down all counters on white squares.) If the number is negative, move down all counters of positive value (those on black squares).

3.   Regard both rows as negabinary numbers. Subtract the first row from the second, using the procedure explained in the previous chapter for negabinary subtraction.

4.   Clear the bottom row by negabinary rules.

   To convert a negabinary number to a signed binary

1.   Express the number in negabinary on row 2.

2.   If the number is positive (an odd number of digits), move down all the negative counters (white squares). If the number is negative (an even number of digits), move down all positive counters (black squares).

3.   Regard both rows as binary numbers. Subtract the first row from the second, using a binary procedure.

4.   Clear the answer by binary rules and prefix the proper sign (plus if the original number was positive, minus if it was negative).

   The answer to the question about negabinary palindromes is that the smallest composite number that is palindromic in negabinary is 21. Its positive form is 10101; its negative form is 111111.

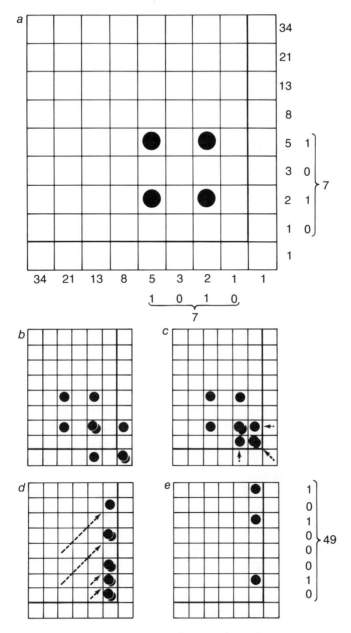

Figure 63 Fibonacci notation for 7 × 7

ADDENDUM

John Harris of Santa Barbara, Calif., discovered an ingenious way to multiply numbers in Fibonacci notation, using the Napier counting board. He added an extra 1-row and 1-column outside the heavy line to the counting board *[see Figure 63]*. Suppose you want to multiply 7 by 7. Place the counters according to

Napier's rules *[see a]*. More counters are now positioned according to the following rule: On the diagonal that extends down and to the right from each counter, *n*, put a counter on every alternate cell, starting with the cell two cells away from counter *n* [*b*].

Each counter outside the heavy line is moved to the nearest cell inside the line [*c*]. Now move all counters up and to the right along their diagonals to the heavy line [*d*]. Clear the column according to the Fibonacci clearing rules given earlier [*e*]. The counters, reading from the top down, give the correct product in Fibonacci notation. Readers familiar with the Fibonacci series will enjoy proving that Harris's algorithm works. Division by this method, however, seems to be hopelessly complicated.

Napier's abacus furnishes insights into many important combinatorial formulas. For example, in how many ways can you make a selection from *n* different objects? The answer $2^n - 1$ is apparent from the way the columns (or rows) are labeled. Let the eight columns of the standard chessboard be eight objects. Each selection of columns corresponds to a binary number from 1 to 11111111, or 255. That $255 = 2^8 - 1$ is obvious, because adding 1 to it makes the binary number 100000000, or $2^8 = 256$.

Assuming one counter to a cell, we can ask several questions about the number of patterns of a specified kind that can be placed on an $n \times n$ chessboard. How many patterns can be formed in which each nonempty column has its counters on the same rows? Clearly this is the same as asking how many products can be made by multiplying two numbers, each from 1 through $2^n - 1$. How many of these patterns have bilateral symmetry along the main diagonal that passes through the board's lower right corner? This is the same as asking how many squares can be made by squaring a number from 1 through $2^n - 1$. How many patterns can be made with no restrictions whatever? Think of the rows as joined to form one long chain of $n \times n$ cells. Every pattern will be expressed by a binary number from 1 through $2^{(n \times n)} - 1$. If we count the absence of all counters as a pattern, the number of patterns possible is 2^{n^2}.

Donald Knuth called my attention to the entertaining article "Binary Notation," by E. William Phillips in the British publication *Journal of the Institute of Actuaries,* Vol. 67, 1936, pages 187–221. The purpose of the paper is to defend a notation based on 8 as superior to decimal notation. To show how easily numbers can be multiplied when given in binary notation, the author reinvents Napier's abacus without realizing it.

Christopher J. Schultz wrote to propose the following procedures for changing a signed binary number to a negabinary number, and vice versa. In many ways they are simpler than the algorithms I gave.

1.   If the number is positive, check the next-to-rightmost column; if negative, check the rightmost column.

2.   If the column contains a counter, consider the columns to the left of it a complete binary number and add 1 to it, using binary arithmetic and clearing rules.

3.   Move two columns to the left and repeat step 2. Continue in this way through the entire number.

To change a negabinary number to a signed binary number:

1.   Starting at the left, check the next-to-first column.

2.   If the column contains a counter, consider the columns to the left of it a complete binary number and subtract 1 from it, using binary arithmetic and clearing rules.

3.   Move two columns to the right and repeat step 2. If the last column checked is the rightmost column, sign the number negative; otherwise sign it positive.

Many readers suggested ways, which they considered better than the one I gave, for performing division on Napier's abacus and also for dividing and doing square roots in Fibonacci notation. Craige Schensted was inspired by Napier's device to invent a chessboard computer on which many astonishing calculations can be made. The basic idea is to allow the columns to be labeled with the powers of one base and the rows to be labeled with the powers of a different base. Each cell represents the product of its row and column numbers. It would require a long chapter to do justice to the elegant ways Schensted found for using such a board to solve problems that otherwise would be difficult.

I gave 1950 as the date on which papers about negative-base number systems first began to appear. In his *History of Binary and other Nondecimal Numeration [see the bibliography]*, Anton Glaser disclosed that in 1885 Vittorio Grünwald published an article in which he covered all the basic arithmetical operations in a negative-10 system. This is the only reference to negative-base notation known to me prior to 1950.

BIBLIOGRAPHY

On Negative Bases

"On Bases for the Sets of Integers." N. G. De Bruijn in *Publication Mathematics Debrecen,* Vol. 1, 1950, pages 232–242.

"A Look at Base Negative Ten." Richard D. Twaddle in *Mathematics Teacher*, Vol. 56, 1963, pages 88–90.

"Using a Negative Base for Number Notation." Chauncy H. Wells, Jr., in *Mathematics Teacher*, Vol. 56, 1963, pages 91–93.

"Negative Radix Arithmetic." Maurits de Regt in *Computer Design*, Vols. 6 and 7, 1967, 1968.

The Art of Computer Programming, Vol. 2, Seminumerical Algorithms. Donald E. Knuth. Addison-Wesley, 1969, page 171, and exercises on pages 176, 177, and 179.

History of Binary and Other Nondecimal Numeration. Anton Glaser, privately published, 1971.

"Negative Based Number Systems." W. J. Gilbert and R. James Green in *Mathematics Magazine*, Vol. 52, 1979, pages 240–244.

On Fibonacci Notation

Fibonacci and Lucas Numbers. Verner E. Hoggatt, Jr. Houghton Mifflin, 1969, pages 70–71.

"Zeckendorf's Theorem and Some Applications." J. L. Brown, Jr., in *Fibonacci Quarterly*, Vol. 2, 1964, pages 162–168.

"Representations of Natural Numbers as Sums of Generalized Fibonacci Numbers." D. E. Daykin in *Journal of the London Mathematical Society*, Vol. 35, 1960, pages 143–160.

"Generalizations of Zeckendorf's Theorem." Timothy J. Keller in *Fibonacci Quarterly*, Vol. 10, 1972, pages 95–112.

CHAPTER NINE

Sim, Chomp and Race track

New mathematical games of a competitive type, demanding more intellectual skill than luck, continue to proliferate both in the U.S. and abroad. In Britain they have become so popular that a monthly periodical called *Games and Puzzles* was started in 1972 just to keep devotees informed. *Strategy and Tactics* (a bimonthly with offices in New York City) is primarily concerned with games that simulate political or military conflicts, but a column by Sidney Sackson reports on new mathematical games of all kinds. Sackson's book *A Gamut of Games* (Random House, 1969) has a bibliography of more than 200 of the best mathematical board games now on the market.

Simulation games are games that model some aspect of human conflict: war, population growth, pollution, marriage, sex, the stock market, elections, racism, gangsterism — almost anything at all. They are being used as teaching devices, and some notion of how widely can be gained from the fact that a 1973 catalogue, *The Guide to Simulation Games for Education and Training,* by David W. Zuckerman and Robert E. Horn, runs to 500 pages.

We will take a look at three unusual new mathematical games. None requires a special board or equipment; all that is needed are pencil and paper (graph paper for the first game) and (for the third) a supply of counters.

Race Track, virtually unknown in this country, is a truly remarkable simulation of automobile racing. I do not know who invented it. It was called to my attention by Jurg Nievergelt, a computer scientist at the University of Illinois, who picked it up on a recent trip to Switzerland.

The game is played on graph paper. A racetrack wide enough to accommodate a car for each player is drawn on the sheet. The track may be of any length or shape, but to make the game interesting it should be strongly curved [see Figure 64]. Each contestant should have a pencil or pen of a different color. To line up the cars, each player draws a tiny box just below a grid point on the starting line. In the example illustrated the track will take three cars, but for

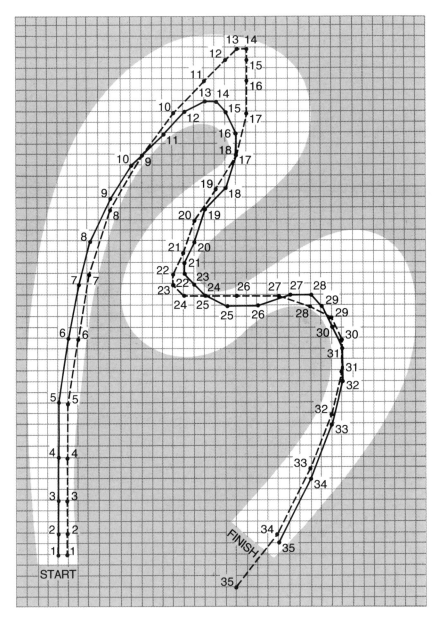

Figure 64 The Race Track game

simplicity a race of two cars is shown. Lots can be drawn to decide the order of moving. In the sample game, provided by Nievergelt, Black moves first.

You might suppose that a randomizing device now comes into play to determine how the cars move, but such is not the case. At each turn a player simply moves his car ahead along the track to a new grid point, subject to the following three rules:

1.   The new grid point and the straight line segment joining it to the preceding grid point must lie entirely within the track.

2.   No two cars may simultaneously occupy the same grid point. In other words, no collisions are allowed. For instance, consider move 22. Gray, the second player, would probably have preferred to go to the spot taken by Black on his 22nd move, but the no-collision rule prevented it.

3.   Acceleration and deceleration are simulated in the following ingenious way. Assume that your previous move was k units vertically and m units horizontally and that your present move is k' vertically and m' horizontally. The absolute difference between k and k' must be either 0 or 1, and the absolute difference between m and m' must be either 0 or 1. In effect, a car can maintain its speed in either direction, or it can change its speed by only one unit distance per move. The first move, following this rule, is one unit horizontally or vertically, or both.

The first car to cross the finish line wins. A car that collides with another car or leaves the track is out of the race. In the sample game Gray slows too late to make the first turn efficiently. He narrowly avoids a crash, and the bad turn forces him to fall behind in the middle of the race. He takes the last curve superbly, however, and he wins by crossing the finish line one move ahead of Black. Neither driver, I should add, always makes his best moves.

Nievergelt programmed Race Track for the University of Illinois's Plato IV computer-assisted instruction system, which uses a new type of graphic display called a plasma panel. Two or three people can play against one another, or one person can play alone. The game became so popular that the authorities made it inaccessible for a week to prevent students from wasting too much time on it.

Our second pencil-and-paper game is called Sim, after Gustavus J. Simmons, a mathematician at the Sandia Corporation laboratories in Albuquerque, who invented it when he was working on his Ph.D. thesis on graph theory. He was not the first to think of it (the idea occurred independently to a number of mathematicians), but he was the first to publish it and to analyze it completely with a computer program. In his note titled "On the Game of Sim" (see the bibliography) he says that one of his colleagues picked the name as short for sImple sImmons, and because the game resembles the familiar game of nim.

Six points are placed on a sheet of paper to mark the vertexes of a regular hexagon. There are 15 ways to draw straight lines connecting a pair of points, producing what is called the complete graph for six points [see Figure 65]. Two

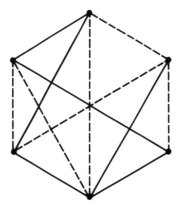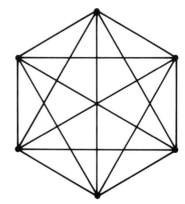

Figure 65   The game of Sim

Sim players take turns drawing one of the 15 edges of the graph, each using a different color. The first player to be forced to form a triangle of his own color (only triangles whose vertexes are among the six starting points count) is the loser.

If only two colors are used for the edges of a chromatic graph, it is not hard to prove that six is the smallest number of points whose complete chromatic graph is certain to contain a triangle with sides all the same color. Simmons gives the proof as follows: "Consider any vertex in a completely filled-in game. Since five lines originate there, at least three must be the same color — say blue. No one of the three lines joining the end points of these lines can be blue if the player is not to form a blue triangle, but then the three interconnecting lines form a red triangle. Hence at least one monochromatic (all one color) triangle must exist, and a drawn game is impossible."

With a bit more work a stronger theorem can be established. There must be at least *two* monochromatic triangles. A detailed proof of this is given by Frank Harary, a University of Michigan graph theorist, in his paper "The Two-Triangle Case of the Acquaintance Graph" *[see the bibliography]*. Harary calls it an acquaintance graph because it provides the solution to an old brainteaser: Of any six people, prove that at least three are mutual acquaintances or at least three are mutual strangers. Harary not only proves that there are at least two such sets but also shows that if there are exactly two, they are of opposite types (colors on the graph) if and only if the two sets have just one person (point) in common.

Because Sim cannot be a draw, it follows that either the first or the second player can always win if he plays correctly. When Simmons wrote his note in 1969, he did not know which player had the win, and in actual play among equally skillful players wins are about equally divided. Later he made an exhaustive computer analysis showing that the second player could always win. Because of symmetry, all first moves are alike. The computer results showed that the second player could respond by coloring any of the remaining 14 edges and still guarantee himself a win. (Actually, for symmetry reasons, there are only two fundamentally different second moves: one that connects with the first move, and one that does not.)

After the first player has made his second move, exactly half of the remaining plays lead to a sure win for the second player and half to a sure loss, assuming, of course, that both sides play rationally. If 14 moves are made without a win, the last move, by the first player, will always produce two monochromatic triangles of his color. This 14-move pattern is unique in the sense that all such patterns are topologically the same. Can you find a way of coloring 14 edges of the Sim graph, seven in one color and seven in another, so that there is no monochromatic triangle on the field?

The most interesting unanswered question about Sim is whether there is a relatively simple strategy by which the second player can win without having to memorize all the correct responses. Even if he has at hand a computer printout of the total game tree, it is of little practical use because it is enormously difficult to locate on the printout a position isomorphic to the one on the board. Simmons's computer results have been verified by programs written by Michael Beeler at the Artificial Intelligence Laboratory of the Massachusetts Institute of Technology and, more recently by Jesse W. Croach, Jr., of West Grove, Pa., but no one has been able to extract from the game tree a useful mnemonic for the second player.

Sim can, of course, be played on other graphs. On complete graphs for three and four points the game is trivial, and for more than six points it becomes too complicated. The pentagonal five-point graph, however, is playable. Although a draw is possible, I am not aware of any proof that a draw is inevitable if both sides make their best moves.

Our third game, which I call Chomp, is a nim-type game invented by David Gale, a mathematician and economist at the University of California at Berkeley. Gale is the inventor of Bridg-it, a popular topological board game still on the market. What follows is based entirely on results provided by Gale.

Chomp can be played with a supply of counters [see Figure 66] or with O's or X's on a sheet of paper. The counters are arranged in a rectangular formation. Two players take turns removing counters as follows. Any counter is selected. Imagine that this counter is inside the vertex of a right angle through the field, the base of the angle extending east below the counter's row and its other side extending vertically north along the left side of the counter's column. All counters inside the right angle are removed. This constitutes a move. It is as though the field were a cracker and a right-angled bite were taken from it by jaws approaching the cracker from the northeast.

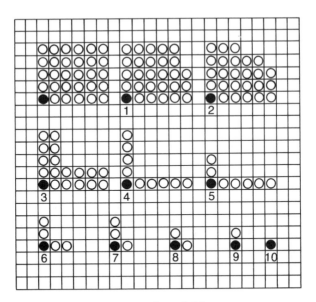

Figure 66   Chomp on a 5-by-6 field

The object of the game is to force your opponent to chomp the poison counter at the lower left corner of the array *[black counter]*. The reverse form of Chomp —winning by taking this counter— is trivial because the first player can always win on his first move by swallowing the entire rectangle.

What is known about this game? First, we dispose of two special cases for which winning strategies have been found.

1.   When the field is square, the first player wins by taking a square bite whose side is one less than that of the original square. This leaves one column and one row, with the poison piece at the vertex *[see Figure 67, left]*. From now on the first player "symmetrizes." Whatever his opponent takes from either line, he takes equally from the other. Eventually the second player must take the poison piece.

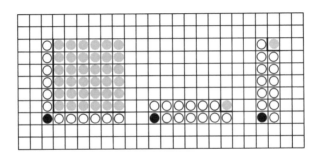

Figure 67   Winning first bites on square field, 2-by-n field and n-by-2 field

2.   When the field is 2 by n, the first player can always win by taking the counter at top right [see Figure 67, middle and right]. Removing that counter leaves a pattern in which the bottom row has one more counter than the top row. From now on the first player always plays to restore this situation. One can easily see that it can always be done and that it ensures a win. The same strategy applies to fields of width 2, except now the first player always makes sure that the left column has one more counter than the right column.

With the exception of these two trivial cases, no general strategy for Chomp is known. There is, however, and this is what makes Chomp so interesting, a simple proof that the first player can always win. Like similar proofs that apply to Bridg-it, Hex, generalized ticktacktoe and many other games, the proof is nonconstructive in that it is of no use in finding a winning line of play. It only tells you that such a line exists. The proof hinges on taking the single counter at the upper right corner in the opening move. There are two possibilities: (1) It is a winning first move; (2) it is a losing first move. If it is a losing one, the second player can respond with a winning move. Put another way, he can take a bite that leaves a position that is a sure loss for the first player. But no matter how the second player bites, it leaves a position that the first player could have left if his first bite had been bigger. Therefore if the second player has a winning response to the opening move of taking the counter at top right, the first player could have won by a different opening move that left exactly the same pattern.

In short, either the first player can always win by taking the counter at top right, or he can always win by some other first move.

"We normally think of nonconstructive proofs in mathematics as being proofs by contradiction," Gale writes. "Note that the above proof is not of that type. We did not start by assuming that the game was a loss for the first player and then obtain a contradiction. We showed directly that there was a winning strategy for the first player. The word 'not' was never used in the argument. Of course we used implicitly the fact that any game of this kind is a win for either the first or the second player, but even the proof of this fact can be given by a simple inductive argument that does not use any law of the excluded middle."

This is essentially all that is known about Chomp except for some curious empirical results Gale obtained from a complete computer analysis of the 3-by-n game for all n's equal to or less than 100. In every case it turned out that the winning first move is unique. Figure 68 shows the winning moves for 3-high fields of widths 2 through 12. Rotating and reflecting these patterns give winning moves on 3-wide fields of heights 2 through 12 because any m-by-n game is symmetrically the same as the n-by-m game.

A winning first move on a 3-high field must be one or two rows deep. (A 3-deep bite would leave a smaller rectangle and thus throw the win to the

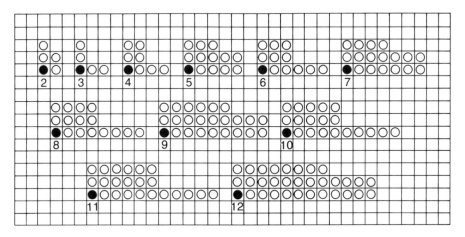

Figure 68   Winning first bites on 3-by-*n* fields

second player.) Roughly 58 percent of the winning first moves are two rows deep, and 42 percent are one row deep. Note that the 1-row moves either stay the same or increase in width as *n* increases, and the same is true of the 2-row moves. A partial analysis of all 3-high fields with widths less than 171 showed that the sole exception to this rule occurs when *n* is 88. The winning first move on the 3-by-88 rectangle is 2 by 36, which is one unit less wide than the winning 2-by-37 move on the 3-by-87 field. "Phenomena like this," Gale writes, "lead one to believe that a simple formula for the winning strategy might be quite hard to come by."

There are two outstanding unproved conjectures:

1.   There is only one winning first move on all fields.

2.   Taking the counter at the top right corner always loses except on 2-by-*n* (or *n*-by-2) fields.

The second conjecture has been established only for fields with widths or heights of 3. Readers are invited to discover the unique winning openings on 4-by-5 and 4-by-6 rectangles.

ANSWERS

Sim has only one basic position (variants are topologically identical) that allows the game to go 14 moves without a monochrome triangle [*see Figure 69, top*]. The 4-by-5 and 4-by-6 fields for Chomp are won by the unique first moves shown at the bottom of Figure 69.

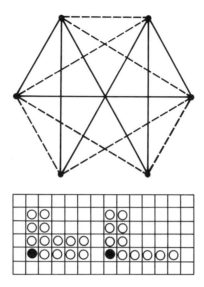

Figure 69   Sim game that ends on move 15 and winning chomps

David Gale, who invented Chomp, has considered the game on infinite rectangular arrays. Readers may enjoy proving (on the basis of the given theorems) that the first player wins on n-by-infinity fields (provided that n is not 2) and on infinity-by-infinity squares but loses on 2-by-infinity arrays.

ADDENDUM

The three games prompted a variety of interesting letters. Many readers felt that Race Track rules should not allow one car to win if another car on the same move could also cross the finish line. They suggested giving the win to the car farthest from the finish line at the end of the move. Joe Crowther was the first of many readers who proposed drawing one or two patches on the roadway to represent oil slicks. Cars are required to move at a constant speed and direction when passing wholly or partly through each patch. J. P. Schell, in addition to oil slicks, proposed adding upgrades and downgrades to force cars to speed up or slow down, as well as stationing pretty girls along the track to distract drivers. Others suggested adding pit stops here and there and requiring a driver to lose one move by coming to zero velocity within any one pit of his choice. Some readers thought it would simplify the game if the finish line were always drawn along one of the grid lines.

David Pope suggested a fast-acceleration move. Whenever a car slows to a full stop, it can, on the next move, go any desired distance in either or both of the two directions. Tom Gordon, who welcomed the game as a teaching device for his high school physics students, added a power-braking option that allows a car to reduce both coordinates by two units, provided the move continues the preceding move in a straight line.

C. R. S. Singleton described two novel variants of the game: (1) Instead of a track, numbered gates are marked on the graph. Cars must pass through the gates in numerical order. (2) A series of numbered checkpoints are substituted for the track. Cars must visit each checkpoint by ending a move on that point.

Michael D. Greenberg and his friends at the Westinghouse Aerospace Division in Baltimore adopted two rules to offset the advantage of a first move: (1) Slant the starting line (as in actual racing) and allow the second player to choose between the two starting points. (2) Allow cars to occupy the same point at the same time. They also preferred to draw the track along grid lines to avoid arguments over whether a point was on the track or inside it. Two British readers, Giles Vaughan-Williams and John Kinory, devised rules allowing cars to brake and skid when rounding a sharp curve at high speed.

I have been unable to determine the origin of Race Track. A car-race game very similar to it appeared under the name of *Le Zip* in a French book by Pierre Berloquin, *Le Livre des Jeux,* published about 1971. It is reprinted in Berloquin's book *100 Jeux de Table* (Paris, 1976). A version of *Race Track* appeared as game 13 in the Hewlett-Packard *Games Pac 1* book (1976) for use with the company's HP-67 and HP-97 calculators.

The game of Sim on a complete graph for five points is now known to be a draw if played rationally. (All draws are topologically equivalent to a pentagram of one color inscribed in a pentagon of the other color. Think of the points as balls connected by elastic strings. If one pattern can be changed to another, they are considered identical.) A complete game tree for five-point Sim was hand-constructed by Eugene A. Herman of Grinnell College and Leslie E. Shader of the University of Wyoming. Jesse W. Croach, Jr., of West Grove, Pa., was able to draw the tree by extracting information from his computer printout for six-point Sim. The first computer program written specifically for five-point Sim was by Ashok K. Chandra of the Artificial Intelligence Laboratory at Stanford University. It produced a complete tree in a few seconds. The results were confirmed by Michael Beeler's program.

Both Chandra and Herman noticed that a good strategy in five-point Sim is to form a closed circuit of four edges of your color, with a fifth edge attached to any of the four dots. This guarantees your win. Herman noticed that as soon as a dot has three edges of the same color attached to it a draw becomes impossible. Variations and generalizations of Sim came from several readers.

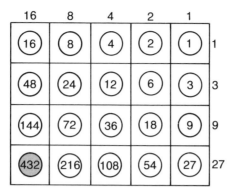

Figure 70   Chomp as a divisor game

The most surprising letter (to put it mildly) was from G. J. Westerink, of Veenendaal in the Netherlands, disclosing that the game of Chomp is isomorphic with a number game invented by the late Fred Schuh, a mathematician at Delft Technical College. It is one of the prettiest isomorphisms I have ever encountered in recreational mathematics. The game does not appear in Schuh's *Master Book of Mathematical Puzzles and Recreations* (Dover, 1968), but he explained it in a 1951 paper cited in the bibliography. Two players agree on any positive integer, N. A list is made of all the divisors (including N and 1); then players take turns crossing out a divisor and all *its* divisors. The person forced to take N loses. Planar Chomp corresponds to this game when N has exactly two prime divisors, solid Chomp to the game when N has three prime divisors, four-dimensional Chomp when N has four prime divisors and so on.

This is best made clear with an example. Consider $N = 432$, a number that prime-factors to $2^4 \times 3^3$. Draw a rectangular Chomp field with sides of 5 and 4 (the exponents raised by 1), and label the four rows with powers of 3 and the five columns with powers of 2. Counters have values that are products of their row and column *[see Figure 70]*. The equivalence of Chomp to the divisor game is now readily apparent. Moreover, any integer whose prime factors have the formula $m^4 \times n^3$ will correspond to the same Chomp field. Incredibly, most of the theorems discovered by David Gale for his game of Chomp (including the beautiful proof of first-player win) had been discovered by Schuh in arithmetical form!

Schuh offered to play readers by correspondence, using $N = 720$. Because the factors of 720 are $2^4 \times 3^2 \times 5^1$, it corresponds to Chomp on a 5-by-3-by-2 field. This proved to have two winning first moves (counters 36 and 48 when numbered according to the system explained). Like Gale, Schuh was unable to find a strategy for first-player win, a way to determine a winning first move short of constructing the game tree or a two-prime (planar Chomp) game that had more than one winning first move.

The first counterexample to the conjecture that all planar games have unique winning first moves was found by Ken Thompson of Bell Laboratories. His computer program produced many examples of fields with two first-move wins, the smallest being 8 by 10. The winning moves leave either five columns of 8 and five of 4, or eight columns of 8 and two of 3. This has been confirmed by Beeler. In 1981 Gil Golani, a student of Z. Wakeman, a mathematician at Ben-Gurion University of the Negev, Israel, wrote a PASCAL program that found a violation of the conjecture on an even smaller board. The two winning first moves for a 6-by-13 rectangle are to take two columns of three lines or five columns of two lines.

Cubical Chomp is an interesting challenge. It is easily seen that the winning first move on the order-2 cube is to take the order-1 cube from the corner. Westerink's analysis of the order-3 cube reveals that the winning first move is to take an order-2 cube from the corner. I previously gave Gale's simple proof that the winning move for any square of order n is to take a square of order $n - 1$. Do winning first moves on all cubes of order n consist in taking a cube of order $n - 1$? If so, does this generalize to n-space cubes?

David Gale reported that in three-dimensional chomp a 2-by-m-by-n game is a trivial win for the first player even when m or n or both are infinite. The first player simply leaves a 2-by-infinity field — a loss for the opponent. The 3 by 3 by 3 and the 3 by 3 by infinity apparently are still unsolved.

In a later letter Gale reported the following result. Suppose the initial field has any finite number of counters in each row but an infinite number in the bottom row. Regardless of the pattern, the game is a win for the first player. Moreover, the winning first move is unique. Gale sent an ingenious nonconstructive proof by contradiction. As in his previous proof of first-player win in standard Chomp, it does not provide what the winning first move is.

Alan Barnert, a Manhattan ophthalmologist and friend, made a right-angled scoop for playing Chomp with a field of raisins. Players ate the raisins scooped on each move until only the poison raisin remained.

David Klarner, in "How to Be a Winner" (see the bibliography), explains how to draw a directed "state graph" that makes visually clear exactly how a first player wins a game of Chomp. Figure 71 shows such a graph for the 2-by-3 game. On the left the actual positions after each move are shown. Arrows indicate possible transitions. On the right is a more abstract diagram of the same graph. Winning states are black. The first player's strategy is to make the initial black move at the top, then to always play to a winning (black) state on his next move. He is certain to reach the final winning state at the bottom. For larger fields, of course, such graphs quickly become too complex to draw.

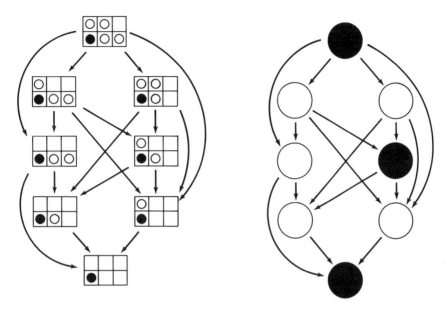

Figure 71   A state graph for 2 × 3 chomp

BIBLIOGRAPHY

Sim

"On Sets of Acquaintances and Strangers at Any Party." A. W. Goodman in *The American Mathematical Monthly*, Vol. 66, 1959, pages 778 – 783.

"SIM — A New Game in Town." Arch Napier in *Empire* (supplement to the *Sunday Denver Post*), May 24, 1970, pages 38 – 40.

"The Two-Triangle Case of the Acquaintance Graph." Frank Harary in *Mathematics Magazine*, Vol. 45, 1972, pages 130 – 135.

"The Game of Sim: A Winning Strategy for the Second Player." E. Mead, A. Rosa and C. Huang *ibid.*, Vol. 47, 1974, pages 243 – 247.

"Another Strategy for Sim." Leslie E. Shader *ibid.*, Vol. 51, 1978, pages 60 – 64.

The following articles appeared in *Journal of Recreational Mathematics:*

1.   "On the Game of Sim." Gustavus J. Simmons, Vol. 2, 1969, page 60.

2.   "Some Investigations into the Game of Sim." A. P. DeLoach, Vol. 4, 1971, pages 36 – 41.

3. "Sim as a Game of Chance." W. W. Funkenbush, Vol. 4, 1971, pages 297–298. A proof that if moves are random the second player has the best chance of winning. Funkenbush has generalized this to *n*-point Sim in an unpublished manuscript.

4. "DIM: Three-Dimensional Sim." Douglas Engel, Vol. 5, 1972, pages 274–275.

5. "Sim on a Desktop Calculator." John H. Nairn and A. B. Sperry, Vol. 6, 1973, pages 243–251.

6. "A Winning Strategy for Sim." E. M. Rounds and S. S. Yau, Vol. 7, 1974, pages 193–202.

7. "The Graph of Positions for the Game of Sim." G. L. O'Brien, Vol. 11, No. 1, 1978–1979, pages 3–9.

8. "Sim with Non-Perfect Players." Benjamin S. Schwartz, Vol. 14, No. 4, 1981–1982, pages 261–265.

All the above articles, with the exceptions of 4 and 8, are reprinted in *Mathematical Solitaires and Games,* edited by Benjamin L. Schwartz. Baywood, 1980. This is an anthology of articles from the *Journal of Recreational Mathematics* issued by the publishers of that journal.

Chomp

"Spel van Delers." (Game of Divisors). Fred Schuh in *Nieuw Tijdschrift voor Wiiskunde* (New Journal of Mathematics), Vol. 39, 1951–1952, pages 299–304.

"A Curious Nim-Type Game." David Gale in *The American Mathematical Monthly,* Vol. 81, 1974, pages 876–879.

"Schuh's 'Spel van Delers' en Gale's 'Chomp.'" G. J. Westerink in *Nieuw Tijdschrift voor Wiiskunde,* Vol. 63, 1975, pages 18–27.

"Chomp for Basic," by Peter Lynn Sessions, and "Chomp for 8008," by Phil Mork in *People's Computer Company,* Vol. 4, 1975, page 10.

"How to Be a Winner." David Klarner, in *Schema,* Vol. 1, No. 2, fall 1981, pages 22–28. This was a short-lived quarterly devoted to mathematical games, published by Michael Waitsman, of Chicago. Only two issues appeared.

CHAPTER TEN

Elevators

Elevators, unlike cars, trains, planes, ships and other common modes of transportation, have been unduly neglected by recreational mathematicians. In this chapter we undertake to rectify the situation by considering four unusual elevator problems. The first three were provided by Donald E. Knuth, a computer scientist at Stanford University and author of a classic seven-volume work in progress titled *The Art of Computer Programming*. Before discussing two combinatorial problems that appear for the first time in his third volume, we consider a well-known probability paradox with a startling generalization that Knuth discovered a few years ago.

George Gamow and Marvin Stern introduced the elevator paradox in the prologue to their little book *Puzzle-Math* (Viking, 1958). Gamow once had an office on the second floor of a seven-story building in San Diego, and Stern had an office on the sixth floor. When Gamow wanted to go up to see Stern, he noticed that in about five cases out of six the first elevator to stop on his floor was going down. It seemed as if elevators were being manufactured on the roof and then sent down the shafts to be stored in the basement. For Stern the situation was the opposite. When he wanted to go down to see Gamow, about five times out of six the first elevator to arrive was on its way up. Were elevators being fabricated in the basement and then sent to the roof to be carried off by helicopters?

The explanation, as Knuth pointed out later, requires a few idealizing assumptions. Suppose each elevator travels independently in continuous cycles from bottom floor to top and back again, moving with constant speed and with the same average waiting time on each floor. Thus at the time a button is pushed on any floor, we can assume that each elevator is at a random point in its cycle.

For a single elevator, calculating the probability that it is on its way down when it stops on a given floor is quite easy. Stern, on the sixth floor, has five floors below and one above; therefore the probability is 5/6 that the elevator is below him and will be moving up when it arrives. Gamow, on the second floor, has five floors above and one below; therefore the probability is 5/6 that the elevator is above him and will reach his floor on its way down. Gamow and Stern explained this in their book, but then they made a slip. If there is more than one elevator, they wrote, the probabilities "of course remain the same." The slip is understandable because the statement seems so intuitively true. Apparently Knuth was the first to realize that it is not true at all! Indeed, as the number of elevators approaches infinity the probability that the first elevator to stop on any floor (except the top or bottom floors) is going up (or down) approaches exactly 1/2 — a rather unexpected result. Yet the probability (for, say, the second floor) remains 5/6 for every individual elevator, and all elevators are equally likely to be the next to arrive.

The solution for two or more elevators is complicated by conditional probabilities. As Knuth puts it: "The choice of which elevator is first to arrive on the second floor is partly contingent on whether it was above us or below, since an elevator that is below the second floor when we begin to wait is likely to arrive ahead of an elevator that is above (all other things being equal)." In his 1969 paper [see the bibliography], Knuth analyzes Gamow's situation as follows: Consider the portion of an elevator's route that starts at the fourth floor, then goes down to the first floor and up to the second, a total of $4/12 = 1/3$ of the entire route. During the first half of this portion the elevator stops next at the second floor going down, and during the other half it will next stop going up. Therefore we may call it the unbiased portion, since it is not biased toward up or down.

If there are n elevators, Knuth now distinguishes two cases:

1.   No elevator is in the unbiased portion. The probability of this is $(2/3)^n$, since it is 2/3 for each elevator. The next elevator to stop on the second floor will be going down.

2.   At least one elevator is in the unbiased portion. The probability is $1 - (2/3)^n$. We can ignore any elevator outside the unbiased portion, since one of those in the unbiased portion will necessarily reach the second floor first. In this case the elevator will be going down with probability 1/2.

Combining these results gives a probability of $(2/3)^n + \frac{1}{2}(1 - (2/3)^n) = \frac{1}{2} + \frac{1}{2}(2/3)^n$ that the first elevator to arrive on the second floor will be going down. If there are just two elevators running in Gamow's seven-story building, the first elevator to stop at the second floor will be headed downward with probability $\frac{1}{2} + 2/9 = 13/18$. This is slightly less than 5/6, so Gamow's chances of catching an up elevator have improved. If there are seven elevators, the probability of an elevator's going down would be $2,315/4,374$, which is not far from 1/2.

Knuth gives the general formula for any building by defining p as the distance from a given floor to the bottom divided by the distance between top and bottom floors. For Gamow p is $1/6$; for Stern p is $5/6$. The general formula for all values of p between 0 and 1 is

$$1/2 + 1/2(1 - 2p)|1 - 2p|^{n-1}.$$

The pair of vertical lines indicates the absolute value of the expression between them. The probability approaches $1/2$ as n, the number of elevators, approaches infinity.

Our second elevator problem is from the third volume of Knuth's series, a book that deals entirely with computer techniques of sorting and searching for information. Like its two predecessors, it is comprehensive in scope, written in a clear, informal style (although at times it is necessarily terse and technical) and rich in humor, historical data and problems of great recreational interest. On pages 11 through 72, for instance, Knuth brilliantly summarizes almost everything known about the combinatorial properties of permutations, a topic that ties in with scores of classic puzzle problems. The book's exercises concern such entertaining topics as solitaire card games, shuffling, anagrams, snowplows, the design of tennis tournaments (including Lewis Carroll's flawed efforts to find a design that does the best possible justice to the second-best player), rook problems, sorting puzzles, the unsolved weight-ranking problem, the Josephus problem, parking problems, Fibonacci numbers, the "tableaux" of Alfred Young (which have a curious relevance to the eightfold way of particle theory) and a hundred other things that lead straight into recreational mathematics.

Here we are concerned with pages 357 through 360, where Knuth regards the elevator as a model of one-tape computer sorting. A building has n floors, each holding exactly c people. There is a single elevator that carries at most b people. We assume that the building is full (contains cn people). Exactly c persons want to go to each floor: c to the first floor, c to the second floor and so on. Some people may already be on their desired floor, but it is more interesting to assume that all or most are misfits who want to be on another floor.

The elevator always starts at the bottom. It moves up and down, loading and unloading passengers, until each person is where he wants to be. The elevator then returns to the first floor. A movement of the elevator from any floor to the next floor above or below will be called a unit trip. The problem is to find an algorithm that will sort all the people in a minimum number of unit trips. This operation is equivalent, of course, to minimizing the distance traveled or (assuming a constant elevator speed) to minimizing the time required for the sorting.

As Knuth points out, the people correspond to records that are to be computer-sorted. The building is the tape, the floors are blocks on the tape and the elevator is the computer memory. A computer can do such things as duplicate records or chop them into parts to be stored temporarily in different blocks. It turns out, however, that a clever algorithm discovered by Richard M. Karp enables the elevator to do its job with peak efficiency without having to duplicate or partition any passenger.

Let k be the number of the floor, u_k the number of misfits on k and all lower floors who want to go higher than k, and d_k the number of misfits on k and all higher floors who want to go lower than k. It is not hard to see that $u_k = d_{k+1}$. For example, suppose k equals 3. The theorem states that all people on Floors 3, 2 and 1 who want to go above Floor 3 are equal in number to those on Floor 4 and higher who want to go below Floor 4. (It is like the old wine-and-water problem. In a filled building the people who go up from a bottom portion of the building must be replaced by the same number of people in the top portion who want to go down.) Both u_n (misfits on the top floor) and d_1 (misfits on the first floor) are, of course, zero, since no one wants to go above the top floor or below the first floor.

Because the elevator holds at most b people, it must make at least $\lceil u_k/b \rceil$ trips from Floor k to the next floor above, where $\lceil \ \rceil$ symbolizes the roundup function (the value is rounded up to the nearest integer). Similarly, the elevator must make at least $\lceil d_k/b \rceil$ trips from k down to the next floor below. If we now calculate $\lceil u_k/b \rceil$ and $\lceil d_k/b \rceil$ for each floor, the sum of all these integers will be the least number of trips that the elevator must make to sort everyone.

Karp's algorithm achieves this minimum if u_k is not zero for any floor except the top one and provided that the number of people each floor can hold is not less than the number the elevator can hold. The procedure calls for the assumption that the elevator is always in either the UP state or the DOWN state. It starts in the UP state and repeats the following algorithm until everyone is sorted:

1.   When the elevator is in the UP state, if anyone (in the elevator or on the floor where it has just stopped) wants to go up, fill the elevator with those of the highest destination, with all others remaining on the floor, then move the elevator up one floor. Otherwise, change to the DOWN state.

2.   When the elevator is in the DOWN state, fill it with those people of the lowest destination (who are on the elevator or on the current floor) and move the elevator down one floor. Then change the elevator to the UP state if there are no misfits on lower floors who want to go to the new current floor or higher.

To see exactly how this operates, consider a five-floor problem *[see Figure 72]*. Each floor holds three people. Each person is represented by a numeral that

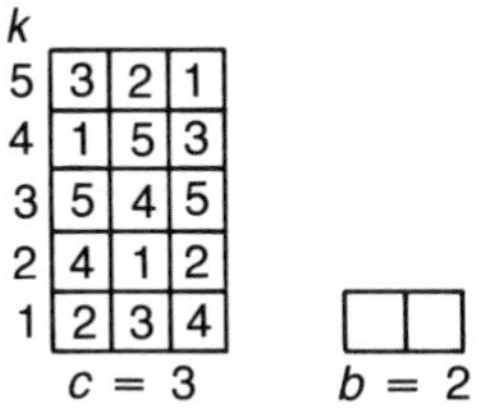

k

5	3	2	1
4	1	5	3
3	5	4	5
2	4	1	2
1	2	3	4

$c = 3$ $b = 2$

u_k/b	d_k/b	$[u_k/b]$	$[d_k/b]$
0	3/2	0	2
3/2	5/2	2	3
5/2	3/2	3	2
3/2	3/2	2	2
3/2	0	2	0
—		—	
9	+	9	= 18

Figure 72   Five-story-building elevator problem

indicates the floor he wishes to go to. The empty elevator on the right can hold only two people. In this problem all the people in the building are misfits except for one 2-person who is already on Floor 2. In order to calculate the minimum distance the elevator must travel, first list the u_k/b and d_k/b values for each floor and then list these values rounded up [see illustration]. Note the positions of the zeros and the fact that the sequence of values for u_k/b is repeated in the d_k/b column except that the sequence starts one floor higher. This repetition is true of all such charts and is a consequence of the theorem $u_k = d_{k+1}$. The sum of the rounded-up values is 18, so we know the elevator must make at least 18 unit trips to accomplish the sorting and then return to the first floor.

Figure 73 shows what happens when we apply Karp's algorithm. (The final step is not shown.) Observe that occasionally people are taken off the floor on which they wish to stay. In some cases the procedure will carry a person in one direction when he wants to go in the other. "This represents," Knuth writes, "their sacrifice to the common good."

To get a feeling for the spooky way Karp's algorithm does its job, readers are urged to work out the problem of sorting 45 people in a nine-floor building with an elevator that holds three people [see Figure 74]. First calculate the minimum number of unit steps needed. Then draw the building on a sheet of cardboard, fill the rooms with small cardboard counters bearing the proper numerals and see how easy it is to apply Karp's algorithm to achieve the minimum. Of course, you can make up endless similar tasks, altering the variables k, c and b as you please and shuffling the people any way you like in the building.

If one or more floors have $u_k = 0$ (that is, no one on that floor or below wants to go above that floor), yet some higher floor has $u_k > 0$, the building becomes divided into disconnected regions. The minimum is achieved by handling each region separately according to Karp's algorithm and then piecing together the individual schedules. This procedure increases the number of unit trips by twice the number of floors that must be passed even though they have $u_k = 0$. A little experimenting on buildings with one or more $u_k = 0$ floors below the top one will make clear why that is so. It amounts to the fact that the elevator has to

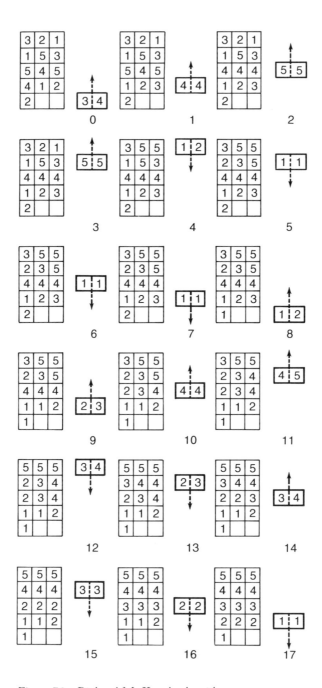

Figure 73   Richard M. Karp's algorithm

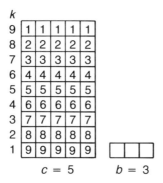

Figure 74 Nine-story building elevator problem

make special trips upward to take care of all higher disconnected regions and then return to the bottom.

Our third elevator problem, discussed on pages 374 through 376 in Knuth's third volume, is based on results obtained by Robert W. Floyd while he was working on efficient ways to rearrange records in a magnetic-disk file. This time instead of minimizing distance we want to minimize the number of stops required by the elevator to complete the sorting. Floyd was able to establish a nontrivial lower bound, but no general algorithm is known that achieves the best possible results, except of course a brute-force trial of all possible elevator schedules.

Consider a building where the number of floors, the number of people to a floor and the elevator capacity are each six *[see Figure 75]*. One of Knuth's exercises is to sort the 36 people correctly by starting and ending the elevator on the first floor and to do it in no more than 12 stops. I shall give Floyd's solution in the answer section.

Floyd's method of computing the lower bound is too complicated to explain here, but for this problem it gives 10 stops. Even in this simple case it is not

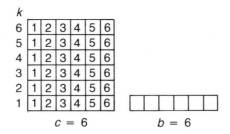

Figure 75 Robert W. Floyd's elevator problem

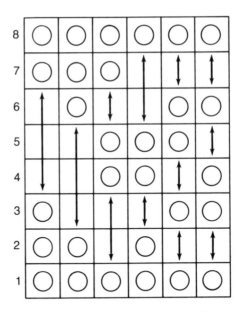

Figure 76   A Japanese elevator problem

known whether there is a solution in 10 or 11 stops. The initial position is not, of course, considered a "stop," but the final move to the first floor is.

Our final problem is from Kobon Fujimura's latest Japanese puzzle book, *Dialogue about Puzzles* (Tokyo, 1971; there is no English translation), which he coauthored with Michio Matsuda. Chapter 3 is devoted to an elevator problem that is a cleverly disguised form of a well-known problem in coding theory. In a building of k floors there are n elevators. Each stops on the top and bottom floors and on exactly m floors in between (always stopping on the same m floors). We wish to determine the minimum number of elevators that will enable a person to go from any floor to any other without changing elevators. For example, suppose a building has eight floors and each elevator stops on top and bottom floors and three floors in between. One schedule for a minimum of six elevators that makes it possible for a person to go directly from any floor to any other floor is shown in Figure 76.

As an introduction to this class of problems, readers are asked to answer the following question. Each elevator in a 10-floor building stops on top and bottom floors and four floors in between. What is the minimum number of elevators that will enable a person to ride from any floor to any other without changing elevators?

ANSWERS

Richard Karp's elevator algorithm requires a minimum of 72 unit trips to sort the 45 people in the nine-floor building. The solution to the second elevator

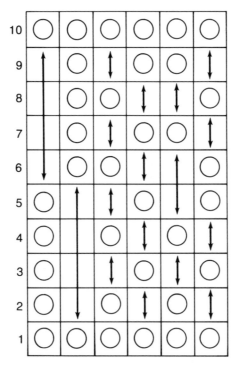

Figure 77 An answer to the elevator problem

problem is given as the answer to Exercise 16 in section 5.4.9 of Donald E. Knuth's *The Art of Computer Programming: Vol. 3, Sorting and Searching.* The best solution known requires 12 stops as follows:

1. 123456 to 2
2. 112334 to 3
3. 224456 to 4
4. 135566 to 5
5. 112334 to 6
6. 224456 to 2
7. 222444 to 4
8. 222222 to 2
9. 555666 to 5
10. 666666 to 6
11. 111333 to 3
12. 111111 to 1

The solution to Kobon Fujimura's elevator problem, a disguised version of an old combinatorial problem of block design, is shown in Figure 77.

ADDENDUM

Robert Floyd's elevator problem was solved in 11 steps by such a large number of readers that listing names is impossible. Most solvers proved 11 to be minimal.

A sample 11-move solution (it is not unique) is 23456 to Floor 2, 33445 to 3, 444556 to 4, 255566 to 5, 122666 to 2, 566666 to 6, 123455 to 5, 123344 to 4, 112333 to 3, 11122 to 2, 11111 to 1.

Solomon W. Golomb was the first to inform me (by telephone) of an 11-step solution. Later he sent a 14-page typescript on the general problem when floors, number of people per floor and elevator capacity all equal k. As he (and others) proved, $2k - 2$ steps (minimal even with unlimited elevator capacity) can be achieved only if k is less than 5. For k greater than 4, $2k - 1$ is the lower bound. Floyd's lower bound shows that $2k - 1$ is impossible if k is 14 or greater.

Allen J. Schwenk wrote to explain how Kobon's problem could be solved by translating it into graph theory. The problem is represented by a complete graph of k points, where k is the number of floors between top and bottom floors. An elevator that stops at m of these floors specifies a subgraph. Schwenk showed that a lower bound for the minimum number of elevators that solve a problem of this kind is given by the formula

$$\left\lceil \frac{k}{m} \left\lceil \frac{k-1}{m-1} \right\rceil \right\rceil$$

Unfortunately, the lower bound is not always obtainable. For example, if $k = 7$ (nine floors altogether) and $m = 4$, the formula gives 4 as the lower bound, but actually five elevators are necessary. It seems likely, as pointed out in the bibliography's 1975 reference (*Journal of Recreational Mathematics*), the number of elevators is either given by the formula, or one more elevator is required.

BIBLIOGRAPHY

"The Gamow-Stern Elevator Problem." Donald E. Knuth in *Journal of Recreational Mathematics*, Vol. 2, 1969, pages 131–137.

The Art of Computer Programming: Vol. 3, Sorting and Searching. Donald E. Knuth. Addison-Wesley, 1973.

"An Elevator Problem." Kobon Fujimura in *Journal of Recreational Mathematics*, Vol. 8, 1975, pages 54–56.

CHAPTER ELEVEN

Crossing Numbers

Modern graph theory is raising many curious questions that appear to be simple but turn out to be extraordinarily complex. An entertaining class of such problems, some the basis of classic puzzles, are those that have to do with "crossing numbers." As Paul Erdös and Richard K. Guy wrote in "Crossing Number Problems" (see the bibliography), "Almost all questions that one can ask about crossing numbers remain unsolved."

Before explaining what a crossing number is, a few fundamental terms must be defined. A graph is a figure consisting of points and lines connecting some of the points. The points are called nodes (or vertices), and the lines are called edges (or arcs). Only the graph's topological structure is significant. Think of the nodes as little spheres joined by elastic strings. Two graphs may look quite different, but if they represent two ways of placing the same ball-and-string model on a surface, they are considered identical.

Where two edges intersect at a point other than their nodes, the point in common is called a crossing. A graph can always be drawn so that no edge crosses itself or crosses an edge joined to one of its nodes, and so that no more than two edges go through any one crossing. Such a drawing is called a "good" drawing. Put another way, a good drawing is one in which each crossing involves two lines that join a distinct set of four points. When a good drawing is designed so that the number of crossings is as small as possible, the minimum number is called the crossing number of that graph.

To make this clearer, consider what is called the complete graph for *n* points. This is a graph on which every pair of nodes are joined by one edge. It is obvious that the crossing number for complete graphs of one, two and three points is 0, and it takes only a moment of pencil doodling to find that it is also 0 for four points. A graph with a crossing number of 0 is called a planar graph.

The simplest nonplanar graph is the complete graph for five points. It has a crossing number of 1. This means that, try as you will, you cannot join all pairs

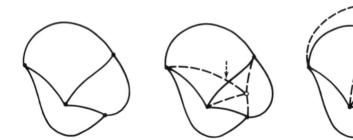

Figure 78   Proof that the complete five-point graph has a crossing number of 1

of the five nodes without producing at least one crossing. This can be proved informally as follows. All forms of the complete graph for four points consist of three mutually contiguous regions [*see Figure 78*]. A fifth point (shown as a circle) must go either inside one of the three regions or outside the entire figure. When the fifth point is inside, you cannot connect it to the node outside its region without crossing an edge. When the fifth point is outside, you cannot get from it to the interior node without crossing an edge. The crossing is indicated by the arrow. (For a different proof see page 94 of *Graphs and Their Uses*, Random House, 1963, by Oystein Ore, an excellent introduction to graph theory.)

The fact that the complete graph for five points is not planar establishes that a map of five regions cannot be drawn so that every pair of regions share a boundary. If such a map could be drawn, we could put a point inside each region, connect each pair of points by an edge that crosses the border shared by the two regions containing those points and do this without creating crossings. In other words, we would be able to draw a complete graph of five points with a crossing number of 0. As we have seen, that is impossible. Unfortunately this does not prove the famous 4-color map theorem.

It is true that for any map, say of many hundreds of regions, any specified set of five regions can always be colored with four colors without having two adjacent regions of the same color. Until 1976 it was conceivable, however, that five colors might still be required for the entire map. If you tried to color it with four, there could always be a place where you ran into trouble. If you eliminated the trouble at that place by recoloring regions, the trouble would pop up at some other spot. That five regions cannot mutually touch was established long ago, but the 4-color map theorem, an altogether different matter, was not solved until 1976.

One might suppose it would be simple to write a formula for the crossing number of a complete graph of *n* points, but this is unsolved. In 1960, writing in

Nabla (the Bulletin of the Malayan Mathematical Society), Guy conjectured that the formula is

$$\frac{1}{4}\left[\frac{n}{2}\right]\left[\frac{n-1}{2}\right]\left[\frac{n-2}{2}\right]\left[\frac{n-3}{2}\right]$$

The brackets indicate that the number inside is rounded down to the nearest integer. It has been proved that the crossing number cannot exceed the value of this expression, but the formula has been verified as exact only for *n* through 10.

If we divide the number of points into odd and even, we can express Guy's formula in more conventional ways. For *n* even the formula is

$$\frac{n(n-2)^2(n-4)}{64}$$

For *n* odd it is

$$\frac{(n-1)^2(n-3)^2}{64}$$

Complete graphs for six and seven points, with crossing numbers of 3 and 9 respectively, are shown in the bottom of Figure 79. The graph for six points (like that for five) is unique, but there are six variations of the seven-point graph. They are dissimilar in the sense that, if you regard the graph as being embedded in the plane, you cannot change one to the other (in terms of our ball-and-string model) without lifting a ball off the plane to carry it over an edge or a node.

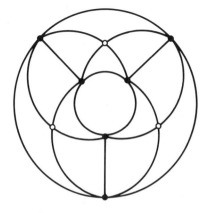 

Figure 79   Six-point graph (left) and seven-point graph (right)

Complete graphs of eight, nine and ten points are known to have crossing numbers of 18, 36 and 60 respectively, as given by Guy's elegant formula. The eight-point graph has three variants. The number jumps to 411 for the nine-point graph, then goes down to 37 for the 10-point graph. Note the curious fact that when n is odd the number of variants is much larger than when n is even, a feature that continues for all higher n's.

An interesting side question occurred to Guy. Can a complete graph with the minimum number of crossing points always be drawn by restricting the edges to straight-line segments? He found that the answer is yes for seven or fewer points and also for nine points, but for eight points the rectilinear crossing number (as it is called when all edges are straight) is 19, not 18. Little is known about rectilinear crossing numbers for complete graphs of more than nine points, although it has been proved that for 10 or more points the rectilinear crossing number is greater than the crossing number. It has been conjectured that the 10-point graph has a rectilinear crossing number of 62.

Here is a pleasant little problem for which a very simple polynomial formula is readily available: What is the maximum number of edges that can be drawn, as part of a complete graph for n points, without a crossing? (Example: For the six-point graph the maximum is 12.)

A formula for the crossing number of complete bigraphs (or bipartite graphs) of m and n points also has not yet been discovered. Such a graph has each point in set m joined to each point in set n, but no edges connect an m point to an m point or an n point to an n point. Complete bigraphs with points of 1,1, 1,2, 2,2 and 2,3 have crossing numbers of 0. The 3,3 graph, known as the Thomsen graph, has a crossing number of 1.

Students of recreational mathematics will at once recognize the 3,3 case as the old "utilities puzzle," so called because Henry Ernest Dudeney presented it with the following story line. There are three houses and three utility sources: water, gas and electricity. The puzzle is to draw lines connecting each house to each utility without any crossings. It cannot be done because the crossing number of this graph is 1 [see Figure 80].

The best conjecture (made by K. Zarankiewicz in 1954) for the crossing-number formula of a complete bigraph is

$$\left[\frac{n}{2}\right]\left[\frac{n-1}{2}\right]\left[\frac{m}{2}\right]\left[\frac{m-1}{2}\right]$$

As with the previous formula, the brackets indicate rounding down, and it has been demonstrated by Zarankiewicz that the crossing number is equal to or less than the number given by the formula. The formula's accuracy has been

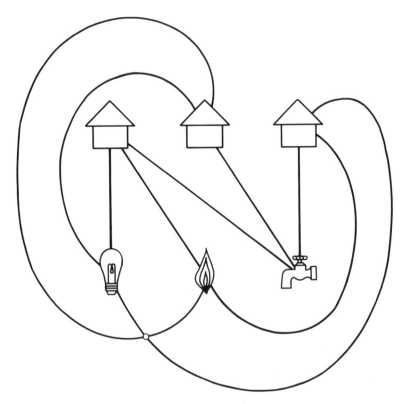

Figure 80 The utilities problem

established (by Daniel J. Kleitman) only for values of m and n through 6. The crossing number for the 7,7 bigraph is not known. Using Zarankiewicz's formula and other arguments, Kleitman proved that this number must be 77, 79 or 81. He ends his 1970 paper with the one-word sentence "Which?"

A rectilinear graph, from the article by Guy and Erdös mentioned earlier, for the 7,7 case with 81 crossings is shown in Figure 81. So far this construction method (in which each set of points is arranged in a straight line, and the two lines are perpendicular) has produced rectilinear graphs that give the lowest crossing numbers known. No one has yet proved that it always does so, although Kleitman told me he believes it does.

As in the case of complete graphs, it is not hard to find a simple polynomial expression for the maximum number of edges that can be drawn, as part of a complete bigraph of m,n points, without a crossing. (Example: For the 3,3 graph the maximum is 8.) Can the reader give the formula?

Some recent work has been done on the crossing numbers of other types of graphs, notably graphs on the plane for the skeletons of n-dimensional cubes,

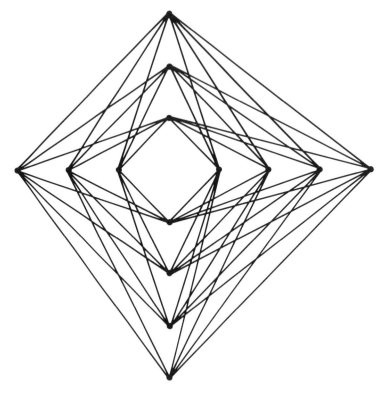

Figure 81   The complete 7,7 bipartite rectilinear graph (81 crossings)

and complete graphs and complete bigraphs on such surfaces as the torus, the Klein bottle and the projective plane. (Graphs drawn on a sphere are the same as those drawn on a plane, because the sphere can be punctured at any spot not on the graph and flattened to a plane without altering the graph's topological structure.)

Richard K. Guy, Tom Jenkyns and Jonathan Schaer, in their paper "The Toroidal Crossing Number of the Complete Graph," prove that the toroidal crossing numbers for seven, eight, nine and ten points are 0, 4, 9 and 23 respectively. (The zero crossing number for seven points on the torus corresponds to the fact that a maximum of seven regions, mutually bordering, can be drawn on the torus.) For 11 points, 42 is strongly believed to be the toroidal crossing number. The best results known for 12, 13, 14, 15 and 16 points are 70, 105, 154, 226 and 326 respectively. The paper gives upper and lower bounds for n greater than 9.

Guy and Jenkyns, writing on toroidal crossing numbers for complete bigraphs, give upper and lower bounds for sufficiently large m and n. The toroidal

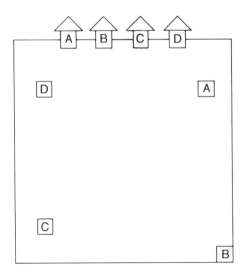

Figure 82   The four-schoolhouses problem

crossing number for a 3,3 graph is 0, which means that the utilities graph is solvable on the torus. Indeed, the authors prove that it can be solved on the torus even if there are four houses and four utilities. The toroidal crossing number is also 0 for graphs of 3,4, 3,5 and 3,6. Complete bigraphs of 4,5, 5,5, 5,6 and 6,6 have toroidal crossing numbers of 2, 5, 8 and 12 respectively. An interesting classroom project is to find ways of drawing these graphs, and those in the preceding paragraph, on the surface of a large model of a doughnut.

Old puzzle books contain many problems based on crossing numbers. Here is an easy problem from one of Dudeney's books. Four boys lived in four houses and went to four schools. Show how the boy in house *A* can walk to school *A*, boy *B* to school *B*, boy *C* to school *C* and boy *D* to school *D* without any of their paths crossing one another or going outside the large square boundary *[see Figure 82]*. Of course, there must be no tricks such as running a path through a house or a school.

ANSWERS

The formula for the maximum number of noncrossing edges that can be drawn as part of a complete graph for n points is $3(n - 2)$ for n greater than 2. The corresponding formula for complete bigraphs of m,n points is $2(m + n - 2)$.

"Odd," a friend once remarked of the bigraph formula, "that the number is always even." Proofs of both cases are not difficult. These formulas for non-crossing edges are of no help in finding formulas for crossing numbers because

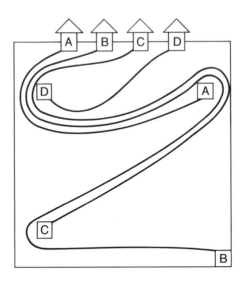

Figure 83   Solution to the schoolhouses problem

there is no known way to predict the minimum number of crossings produced by the edges not drawn.

One solution to the four-schoolhouses puzzle, in which four boys have to reach their respective schools without any of their paths crossing one another or going outside the boundary, is shown in Figure 83.

ADDENDUM

The search for formulas for minimal crossing numbers was initiated in 1944 by the Hungarian mathematician Paul Turán. He was working in a war labor camp, at a brick factory outside Budapest. Here is how he told the story in "A Note of Welcome" in the first issue of *The Journal of Graph Theory,* Vol. 1, 1977, pages 7 – 9:

> There were some kilns where the bricks were made and some open storage yards where the bricks were stored. All the kilns were connected by rail with all the storage yards. The bricks were carried on small wheeled trucks to the storage yards. All we had to do was to put the bricks on the trucks at the kilns, push the trucks to the storage yards, and unload them there. We had a reasonable piece rate for the trucks, and the work itself was not difficult; the trouble was only at the crossings. The trucks generally jumped the rails there, and the bricks fell out of them; in short this caused a lot of trouble and loss of time which was rather precious to all of us (for reasons not to be discussed here). We were all sweating and cursing at such occasions, I too; but *nolens-volens* the idea occurred to me that this loss of

time could have been minimized if the number of crossings of the rails had been minimized. But what is the minimum number of crossings? I realized after several days that the actual situation could have been improved, but the exact solution of the general problem with m kilns and n storage yards seemed to be very difficult and again I postponed my study of it to times when my fears for my family would end. (But the problem occurred to me again not earlier than 1952, at my first visit to Poland where I met Zarankiewicz. I mentioned to him my "brick-factory"-problem.

Turán goes on to tell how Zarankiewicz believed he had solved the bigraph crossing-number problem, but a gap was found in his proof, and the problem became the notorious unsolved question that it remains today.

Roger Baust pointed out in a letter that when the maximum number of noncrossing lines is drawn for a set of points, the graph will consist entirely of regions that are each surrounded by just three points, including the region outside the graph.

Donald Miller sent a generalization of the bigraph (bipartite) problem I asked readers to solve. Instead of two sets of points, consider the "multipartite" graph consisting of k sets of points, where k can take any integer value. No points within a set may be joined, but we wish to connect as many points as possible, without any lines crossing, that belong to different sets. The maximum number of such noncrossing lines is $2(a + b + c + d + . . .) + k - 6$, where $a, b, c, . . .$ are the numbers of points in each set. Thus, for three sets the formula is $2(a + b + c) - 3$. For four sets it is $2(a + b + c + d) - 2$. Note that if a complete graph of n points is viewed as a special case of k sets, each consisting of just one point, the formula reduces to $3(n - 2)$ as previously observed.

The most important practical application of crossing-number theory is in the designing of printed circuits and microchips, where it is desirable to have as few crossings of lines as possible. A more whimsical application is to the spelling of words and phrases. See "Ensnaring the Elusive Eodermdrome," by G. S. Bloom, J. W. Kennedy and P. J. Wexler, and "Dictionary of Eodermdromes" by A. Ross Eckler, both in *Word Ways* (a quarterly journal devoted to word play), Vol. 13, 1980. For example, if the 15 different letters of *supercalifragilisticexpialidocious* are attached to the points of a 15-point graph with no crossings, the word can be spelled by tracing a continuous path from point to point. Eodermdromes are words that cannot be spelled in this way on planar graphs. The challenge is to find graphs on which they can be spelled with as few lines as possible.

David Singer, in an unpublished paper, proved that the crossing number for the rectilinear graph of 10 points must be at least 61, and he constructed such a

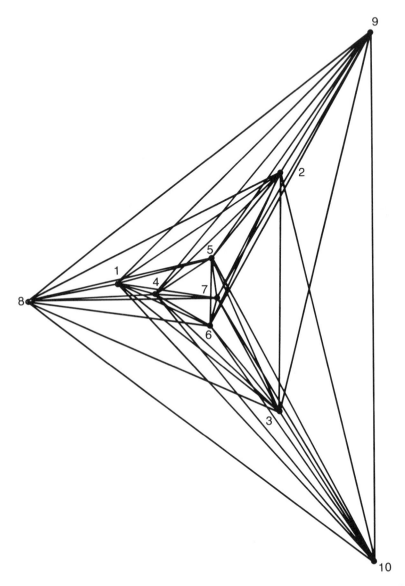

Figure 84   A complete rectilinear graph for 10 points, proving its crossing number can be as low as 62. Is it 61?

graph with 62 crossings *[see Figure 84]*. Guy has pointed out (personal communication) that by examining the 411 known varieties of the complete graph for 9 points it should not be difficult to pin the number down to either 61 or 62, but at this writing it has not been done.

Guy called my attention to two very simple ways of proving that the crossing number is 1 for both the complete graph of five points and the 3,3 bigraph. Both are *reductio ad absurdum* proofs based on Euler's famous formula for maps drawn on a plane or sphere. The sphere case is the easiest to explain. If a map is drawn on a sphere, the regions, lines and points are related by the formula $R = L + 2 - P$, where R stands for regions, L for lines and P for points.

Assume that five points can be joined to make a complete graph with no crossings. By Euler's formula there will be $10 + 2 - 5 = 7$ regions. View each line as a boundary with two "sides," one belonging to each of the two regions separated by the line. There will be $2 \times 10 = 20$ sides. But if there are seven regions, each region will have three sides, making a total of 21. Contradiction! A complete graph for five points, with no crossings, is impossible, and we know such a graph can be constructed with one crossing.

The proof for the 3,3 bigraph is similar. Euler's formula shows there are $9 + 2 - 6 = 5$ regions, assuming the map has no crossings. In this case each region has four "sides," or $4 \times 5 = 20$ sides in all. But twice the number of lines is $2 \times 9 = 18$. Again there is a contradiction, proving that the 3,3 bigraph without a crossing is impossible, and again such a graph can be exhibited with one crossing.

In 1983 Michael Garey and David Johnson, at Bell Laboratories, proved that the problem of calculating the crossing number of a graph (their proof can be extended to cover rectilinear crossing numbers) belongs to a class of problems known as NP-complete. As the number of points of a complete graph increases, the task of calculating the graph's crossing number quickly goes outside the bounds of reasonable computing time. This means there is probably no efficient algorithm that will design microchips or printed circuits, with a large number of impulse-carrying lines, so that the crossings are reduced to the absolute minimum.

BIBLIOGRAPHY

So many papers have been published about crossing numbers during the past few decades that I must content myself with listing only a few highlights and the papers cited in the chapter.

"A Combinatorial Problem." Richard K. Guy in *Nabla*, Vol. 7, 1960, pages 68–72.

"The Toroidal Crossing Number of the Complete Graph." Richard K. Guy, Tom Jenkyns and Jonathan Schaer in *Journal of Combinatorial Theory,* Vol. 4, 1968, pages 376–390.

"The Toroidal Crossing Number of $K_{m,n}$." Richard K. Guy and T. A. Jenkyns in *Journal of Combinatorial Theory,* Vol. 6, 1969, pages 235–250.

"Toward a Theory of Crossing Numbers." W. T. Tutte in *Journal of Combinatorial Theory*, Vol. 8, 1970, pages 45–53.

"The Crossing Number of $K_{5,n}$." Daniel J. Kleitman in *Journal of Combinatorial Theory*, Vol. 9, 1970, pages 315–323.

"Latest Results on Crossing Numbers." Richard K. Guy in *Recent Trends in Graph Theory, Lecture Notes in Mathematics*, Volume 186. Springer-Verlag, 1971.

"Crossing numbers of Graphs." Richard K. Guy, in *Graph Theory and Applications*, edited by Y. Alavi, D. R. Lick, and A. T. White. Springer-Verlag, 1972

"Crossing Number Problems." Paul Erdös and R. K. Guy in *American Mathematical Monthly*, Vol. 80, 1973, pages 52–58.

Crossing Numbers on Graphs. Roger B. Eggleton. Unpublished Ph.D. thesis on work done under Guy at the University of Calgary, Alberta, Canada, 1973. The typescript runs to 189 pages and contains an extensive bibliography.

"Crossing Number is NP-Complete." M. R. Garey and D. S. Johnson in *SIAM Journal on Algebraic and Discrete Methods*, Vol. 4, 1983, pages 312–316.

CHAPTER TWELVE

Point Sets on the Sphere

Two unusual problems involving the surface of a sphere, which only recently have been partly solved, provide an entertaining introduction to some elementary topological properties of point sets. The ingenious partial solutions are not hard to understand. It would, however, be hard to imagine two problems in combinatorial point-set geometry more remote from foreseeable practical applications unless one thinks of recreational mathematics (with its two virtues: amusement and instruction) as a branch of applied mathematics.

The first problem was raised by the Polish mathematician J. G. Mikusiński. (It is Problem 84 in *The New Scottish Book,* a collection of unsolved problems edited by H. Fast and S. Swierczkowski, published in Warsaw in 1958.) Is it possible, Mikusiński asked, to completely cover the surface of a sphere with congruent, nonoverlapping arcs of great circles? The word "congruent" must be carefully defined. As it is used here, it means more than equality of length and curvature. Two great-circle arcs of the same length may differ topologically with respect to their end points. There are three possibilities: An arc is "closed" if it contains both end points, "open" if it excludes both end points and "half-open" if it contains one end point but not the other. Two great-circle arcs are congruent when they are the same length and have the same topological properties.

In a 1964 paper John Horton Conway and Hallard T. Croft, both at the University of Cambridge, proved that the sphere could be covered with congruent arcs of the half-open type and could not be covered with congruent arcs of the open type. Whether or not it can be covered with congruent closed arcs remains unanswered.

Before explaining these results, the authors first solve the analogous problem of covering the Euclidean plane with congruent line segments of the three types. If all the segments are half-open, the answer is obviously yes. Divide the plane into an infinity of horizontal lines. Each line is then filled with half-open

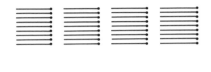

Figure 85   Plane with half-open line segments
Line with closed segments

segments placed head to tail *[see Figure 85, top]*. In this illustration and subsequent ones, a closed end is indicated by a dot. The head of each segment supplies the missing end point at the tail of the next segment.

If the segments are closed, the answer is again yes, but a proof is less trivial. It is impossible to fill a line with closed segments even if they are allowed to vary in length. Figure 85, bottom, shows part of a line divided into closed segments. Because segments are not allowed to overlap, there is no way to join the segments without leaving points uncovered at the joints. Think of the line as the real-number line. Each dot (closed end of a segment) represents a real number. Between any two real numbers, however, there is an uncountable infinity of other real numbers. At every meeting spot of two closed ends an uncountable infinity of points remain uncovered. A meeting of two open ends fails to cover a single point between them, therefore open segments on a line leave a countable infinity of points uncovered. The only way to catch all the points is to have at each joint an open end meet a closed end.

Conway and Croft succeed in covering the Euclidean plane with congruent closed segments by first building a vertical pillar of segments, which goes north and south to infinity *[see 1 in Figure 86]*. On each side they place horizontal pillars *[2 and 3]*, one extending east to infinity, the other west to infinity. Each of these horizontal pillars is open at the finite end, so it joins a side of the vertical pillar without leaving any uncovered points in between. Slanted pillars *[4, 5, 6 and 7]*, open at their finite ends, are added. Eight more slanted pillars go into the gaps *[only two, 8 and 9, are shown]*, then the new gaps are filled with 16 slanted pillars, then 32, and so on, all going to infinity and open at their finite ends. After an infinite number of such steps, every point on the plane will be covered by congruent, nonoverlapping closed segments. The plane cannot be covered with congruent open segments, but the proof is complicated, and readers who want it should refer to the paper by Conway and Croft, in the bibliography.

The authors next turn to Mikusiński's sphere problem. A covering of the sphere by congruent half-open arcs was found by Conway. Assume a unit radius for the sphere. We choose (for a reason that will soon be apparent) half-open arcs of a length that is an integral fraction equal to or less than

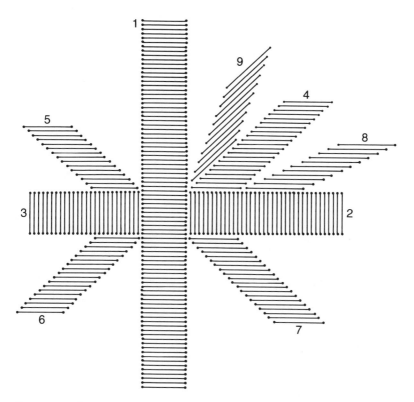

Figure 86   Covering plane with closed segments

one-fifth of a great circle. The first step is to cover the northern hemisphere, except for the north pole and the equator, with such arcs. Let the arcs fan outward from the pole, all with their open tails at the pole, thus leaving the pole uncovered *[see Figure 87]*. The rest of the hemisphere is now divided into an infinite set of thinner and thinner rings, each covered as shown with downward-pointing arcs. It is obvious that we can slant the arcs in these rings so that their slopes approach zero as the rings approach the equator but do not include it. (Does the reader see why this could not be done with arcs equal to onequarter of a great circle?)

The north pole is covered by the following trick. Select a chain of arcs from the pole to the equator, each arc with its head at the tail of the next; then reverse all these arcs so that they point the other way. The top arc covers the pole with its closed end. The entire northern hemisphere is now covered by a cap that is open along its base circle.

A similar procedure caps the southern hemisphere. The final step is to "put a girdle round about the earth" (as Conway and Croft write, quoting Puck) to cover the equator. The girdle is a closed chain of head-to-tail arcs. That is why

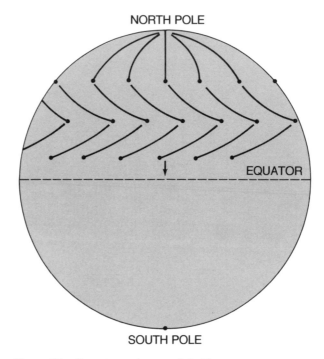

NORTH POLE

EQUATOR

SOUTH POLE

Figure 87   Covering sphere with half-open congruent arcs

the arcs must be integral fractions of a great circle. After the girdling the entire sphere is covered.

For Croft's remarkable proof that the sphere cannot be covered with congruent open arcs, which is too technical to give here, the reader must consult the paper he wrote with Conway. Whether congruent closed arcs can cover the sphere appears to be a difficult question, as yet unsettled. The Conway-Croft paper extends the problem to n-dimensional spaces with some unexpected results. For example, three-dimensional space can be completely filled by the perimeters of nonoverlapping congruent circles.

Our second problem, from Paul Erdös, is unpublished, although Erdös undoubtedly has mentioned it in one of his many lectures on unsolved problems in graph theory. We wish to color all points on the surface of a unit sphere so that no matter how we inscribe an equilateral triangle of side $\sqrt{3}$ (the largest such triangle that can be inscribed) the triangle will have each corner on a different color. What is the minimum number of required colors?

It is easy to prove that six colors are sufficient by coloring the sphere as shown in Figure 88. The polar caps, both open along their boundaries, have parallel base circles whose diameters are $\sqrt{3}$. The rest of the sphere is divided into four congruent regions, each closed along its northern, southern and eastern

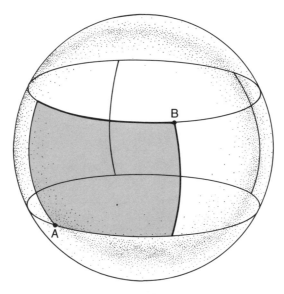

Figure 88   A 6-coloring of a sphere

borders, as shown by the heavy black lines on the perimeter of the dark shaded region; that is, these three borders belong to that colored region. A quick bit of spherical trigonometry shows that points *A* and *B* are less than 120 degrees apart, so they cannot be the vertexes of an inscribed equilateral triangle of $\sqrt{3}$. Each of the six regions is, of course, a different color.

A clever modification of this coloring, by Ernest G. Straus, proves that five colors also are sufficient. Straus's coloring (here published for the first time) is given in a forthcoming paper by Gustavus J. Simmons of the Sandia Corporation in Albuquerque, N.M. The paper [see the bibliography; it was published in 1974] is devoted mainly to Simmons's proof that a 3-color solution to the problem is not possible. Before explaining Simmons's elegant argument, let us see how Straus achieved his 5-coloring *[see Figure 89]*. The north polar region is covered by a cap identical with the north polar cap of the 6-coloring and is also open along its circular base. The rest of the sphere is divided into four identical regions, each closed along its northern and eastern borders, as indicated by the heavy black line on the dark shaded region. One color is given to the cap and to the south pole. Four other colors are assigned to the four quadrant regions.

I now give the proof of 3-color impossibility in Simmons's words. Illustrations for this proof, as well as for the 5- and 6-colorings, were supplied by Simmons.

> "Assume that there is a 3-coloring of the unit sphere satisfying Erdös' problem. Choose any great circle on this sphere and inscribe an equilateral

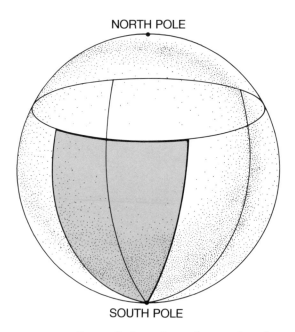

NORTH POLE

SOUTH POLE

Figure 89   Ernest G. Straus's 5-coloring of a sphere

triangle of side $\sqrt{3}$ in it. By hypothesis, the three vertexes are each a different color. Clearly the circle of radius $\sqrt{3}/2$ formed on the surface of the sphere by rotating the equilateral triangle about the axis of symmetry through any vertex must be 2-colored *[see Figure 90, left].*

"Furthermore, every pair of diametrically opposite points on this circle must be of different colors since they are vertexes of an equilateral triangle having the fixed vertex of the generating triangle as a third vertex. All circles of radius $\sqrt{3}/2$ on the unit sphere, which we shall refer to as base circles, have these same properties.

"Choose an arbitrary great circle on the unit sphere and inscribe a Star of David on it *[see Figure 90, right].* Next rotate the equilateral triangles of the star about the axis of symmetry through each vertex in turn to form the six base circles *[see Figure 91].*

"By the symmetry of the construction, it is easy to see that each pair of points connected by [dotted] lines are diametrically opposite each other in one of the base circles and hence must be of different colors. Now, if A has color 1 and B has color 2, then neither G nor H can have color 3. For if G had color 3 (by assertions 1 and 2), J would have color 1 and C and D would both have to have color 3, hence (by assertion 2) E would have color 1 and F color 2. But then K would have to have color 1, which is impossible (by assertion 3) since J also has color 1. On the other hand, if A and B both have the same color 1, then G and H cannot have a different color 2. For if

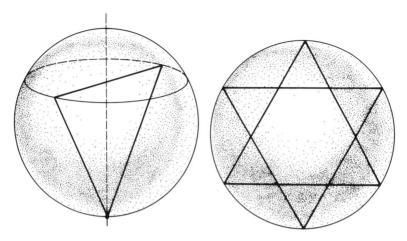

Figure 90   (left) Inscribed and rotated equilateral triangle
(right) Inscribed Star of David

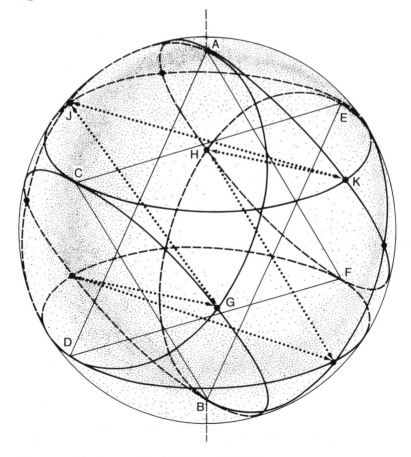

Figure 91   Six base circles from Star of David

G had color 2, say, then C and D would both have to have color 2 also and hence E would be color 3. But then J would have to be either color 2 or 3 (by assertion 1) and at the same time would have to be color 1 (by assertion 2), which is impossible.

"Therefore, by rotating the configuration about the AB axis, it follows that all the points of the equatorial great circle swept out by G and H are colored by the same color or two colors as A and B are. Hence every equilateral triangle inscribed in the equator has two vertexes of the same color. This contradicts the initial assumption and completes the proof that a 3-coloring is not possible."

It still, however, leaves Erdös's problem not fully resolved. We know that five colors are sufficient, and at least four are necessary. How many colors are both necessary and sufficient? No one yet knows whether the answer is 4 or 5.

What about the analogous problem for the plane; that is, what minimum coloring of the plane ensures that every equilateral triangle of unit side will have its corners on different colors? Reflecting on the fact that any two corners of an equilateral triangle are the same distance apart, it seems intuitively clear to us that this is the same as asking for a minimum coloring of the plane so that every unit line segment has its end points on different colors. The intuition is correct. The equivalence of the two problems follows from a beautiful theorem of Erdös and N. G. de Bruijn (1951). It states that a planar graph has a chromatic number k if, and only if, k is the chromatic number for all its finite subgraphs. (The chromatic number of a graph is the minimum number of colors needed for coloring its vertexes so that no two adjacent vertexes are the same color.)

By the same reasoning the sphere problem discussed earlier is the same as asking for a minimum coloring of the unit sphere so that every inscribed line segment of length $\sqrt{3}$ has its ends on different colors. In graph-theory terms, What is the chromatic number of the graph on the sphere in which any two nodes are connected if, and only if, they are separated by an angle of 120 degrees?

The planar problem can be expressed in graph-theory language as, What is the chromatic number of the infinite graph on the plane in which any two points are connected if, and only if, they are a unit distance apart? This question is also due to Erdös, and it too is unsolved. Indeed, the gap between upper and lower bounds is even wider. See Chapter 18 of my *Wheels, Life, and Other Mathematical Amusements* (W. H. Freeman, 1983) for proofs that four colors are necessary, and seven sufficient.

Here is an easy but delightful problem, of unknown origin, that readers may wish to solve. Simmons showed in 1971 [see the bibliography] that if the points of the plane are arbitrarily given a finite number of different colors, there will be

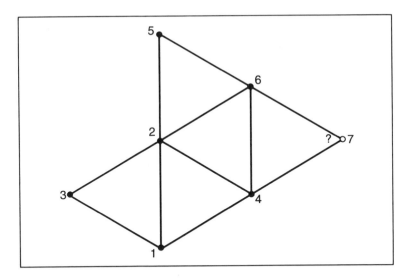

Figure 92   Solution to coloring problem

an uncountable infinity of monochromatic equilateral triangles — that is, triangles with all corners the same color. Long before this was established, however, it was possible to prove a much weaker theorem by using only seven points. The problem is this: Using just seven points, show that if the points on the plane are divided arbitrarily into two differently colored sets, at least one set will contain the vertexes of a monochromatic equilateral triangle.

ANSWERS

Using no more than seven points, prove that if the points of the plane are arbitrarily colored by two colors (say red and black), at least one set will contain the vertexes of a monochromatic triangle. The proof is as follows. Consider any two red points. Call them 1 and 2 and add five more points in a triangular lattice pattern [see Figure 92]. To avoid red equilateral triangles, 3 and 4 must be black. This in turn requires that 5 be red (otherwise triangle 3, 4, 5 is blue), which in turn requires that 6 be black (otherwise triangle 2, 5, 6 is red). Point 7 must be red or black. If red, triangle 1, 5, 7 is red; if black, triangle 4, 6, 7 is black.

BIBLIOGRAPHY

Combinatorial Geometry in the Plane. Hugo Hadwiger and Hans Debrunner. Holt, Rinehart and Winston, 1964.

"Covering a Sphere with Unit Great-Circle Arcs." J. H. Conway and H. T. Croft in *Proceedings of the Cambridge Philosophical Society,* Vol. 60, 1964, pages 787–800.

"Combinatorial Properties of Plane Partitions." G. J. Simmons in *Studia Scientarium Mathematicarum Hungarica,* Vol. 6, 1971, pages 335–339.

"On a Problem of Erdös Concerning a 3-Coloring of the Unit Sphere." G. J. Simmons in *Discrete Mathematics,* Vol. 8, 1974, pages 81–84.

"Bounds on the Chromatic Number of the Sphere." Gustavus J. Simmons in *Proceedings of the Sixth Southeastern Conference on Combinatorics, Graph Theory, and Computing.* Florida Atlantic University, February 17–20, 1975, pages 541–548.

"The Chromatic Number of the Sphere." Gustavus J. Simmons in *Journal of the Australian Mathematical Society Series A,* Vol. 21, 1976, pages 473–480.

CHAPTER THIRTEEN

Newcomb's Paradox

A common opinion prevails that the juice has ages ago been pressed out of the free-will controversy, and that no new champion can do more than warm up stale arguments which every one has heard. This is a radical mistake. I know of no subject less worn out, or in which inventive genius has a better chance of breaking open new ground.

— WILLIAM JAMES

One of the perennial problems of philosophy is how to explain (or explain away) the nature of free will. If the concept is explicated within a framework of determinism, the will ceases to be free in any commonly understood sense, and it is hard to see how fatalism can be avoided. *Che sarà, sarà.* Why work hard for a better future for yourself or for others if what you do must always be what you do do? And how can you blame anyone for anything if he could not have done otherwise?

On the other hand, attempts to explicate will in a framework of indeterminism seem equally futile. If an action is not caused by the previous states of oneself and the world, it is hard to see how to keep the action from being haphazard. The notion that decisions are made by some kind of randomizer in the mind does not provide much support for what is meant by free will either.

Philosophers have never agreed on how to avoid the horns of this dilemma. Even within a particular school there have been sharp disagreements. William James and John Dewey, America's two leading pragmatists, are a case in point. Although Dewey was a valiant defender of democratic freedoms, his metaphysics regarded human behavior as completely determined by what James called the total "push of the past." Free will for Dewey was as illusory as it is in the psychology of B. F. Skinner. In contrast, James was a thoroughgoing indeterminist. He believed that minds had the power to inject genuine novelty into history—that not even God himself could know the future except partially. *"That,"* he wrote, "is what gives the palpitating reality to our moral life and makes it tingle . . . with so strange and elaborate an excitement."

A third approach, pursued in depth by Immanuel Kant, accepts both sides of the controversy as being equally true but incommensurable ways of viewing human behavior. For Kant the situation is something like that pictured in one of Piet Hein's "grooks":

A bit beyond perception's reach
I sometimes believe I see
That Life is two locked boxes, each
Containing the other's key.

Free will is neither fate nor chance. In some unfathomable way it partakes of both. Each is the key to the other. It is not a contradictory concept, like a square triangle, but a paradox that our experience forces on us and whose resolution transcends human thought. That was how Niels Bohr saw it. He found the situation similar to his "principle of complementarity" in quantum mechanics. It is a viewpoint that Einstein, a Spinozist, found distasteful, but many other physicists, J. Robert Oppenheimer for one, found Bohr's viewpoint enormously attractive.

What has free will to do with mathematical games? The answer is that in recent decades philosophers of science have been wrestling with a variety of queer "prediction paradoxes" related to the problem of will. Some of them are best regarded as a game situation. One draws a payoff matrix and tries to determine a player's best strategy, only to find oneself trapped in a maze of bewildering ambiguities about time and causality.

A marvelous example of such a paradox came to light in 1970 in the paper "Newcomb's Problem and Two Principles of Choice" by Robert Nozick, a philosopher at Harvard University. The paradox is so profound, so amusing, so mind-bending, with thinkers so evenly divided into warring camps, that it bids fair to produce a literature vaster than that dealing with the prediction paradox of the unexpected hanging. (See Chapter 1 of my *Unexpected Hanging and Other Mathematical Diversions.*)

Newcomb's paradox is named after its originator, William A. Newcomb, a theoretical physicist at the University of California's Lawrence Livermore Laboratory. (His great-grandfather was the brother of Simon Newcomb, the astronomer.) Newcomb thought of the problem in 1960 while meditating on a famous paradox of game theory called the prisoner's dilemma. A few years later

Newcomb's problem reached Nozick by way of their mutual friend Martin David Kruskal, a Princeton University mathematician. "It is not clear that I am entitled to present this paper," Nozick writes. "It is a beautiful problem. I wish it were mine." Although Nozick could not resolve it, he decided to write it up anyway. His paper appears in *Essays in Honor of Carl G. Hempel,* edited by Nicholas Rescher and published by Humanities Press in 1970. What follows is largely a paraphrase of Nozick's paper.

Two closed boxes, B1 and B2, are on a table. B1 contains $1,000. B2 contains either nothing or $1 million. You do not know which. You have an irrevocable choice between two actions:

1. Take what is in both boxes.

2. Take only what is in B2.

At some time before the test a superior Being has made a prediction about what you will decide. It is not necessary to assume determinism. You only need be persuaded that the Being's predictions are "almost certainly" correct. If you like, you can think of the Being as God, but the paradox is just as strong if you regard the Being as a superior intelligence from another planet, or a supercomputer capable of probing your brain and making highly accurate predictions about your decisions. If the Being expects you to choose both boxes, he has left B2 empty. If he expects you to take only B2, he has put $1 million in it. (If he expects you to randomize your choice by, say, flipping a coin, he has left B2 empty.) In all cases B1 contains $1,000. You understand the situation fully, the Being knows you understand, you know that he knows and so on.

What should you do? Clearly it is not to your advantage to flip a coin, so that you must decide on your own. The paradox lies in the disturbing fact that a strong argument can be made for either decision. Both arguments cannot be right. The problem is to explain why one is wrong.

Let us look first at the argument for taking only B2. You believe the Being is an excellent predictor. If you take both boxes, the Being almost certainly will have anticipated your action and have left B2 empty. You will get only the $1,000 in B1. Contrariwise, if you take only B2, the Being, expecting that, almost certainly will have placed $1 million in it. Clearly it is to your advantage to take only B2.

Convincing? Yes, but the Being made his prediction, say a week ago, and then left. Either he put the $1 million in B2, or he did not. "If the money is already there, it will stay there whatever you choose. It is not going to disappear. If it is not already there, it is not going to suddenly appear if you choose only what is in the second box." It is assumed that no "backward causality" is operating; that is, your present actions cannot influence what the Being did last week. So why not take both boxes and get everything that is there? If B2 is filled, you get $1,001,000. If it is empty, you get at least $1,000. If you are so foolish as to take only B2, you know you cannot get more than $1 million, and there is even a slight possibility of getting nothing. Clearly it is to your advantage to take both boxes!

"I have put this problem to a large number of people, both friends and students in class," writes Nozick. "To almost everyone it is perfectly clear and obvious what should be done. The difficulty is that these people seem to divide almost evenly on the problem, with large numbers thinking that the opposing half is just being silly.

"Given two such compelling opposing arguments, it will not do to rest content with one's belief that one knows what to do. Nor will it do to just repeat one of the arguments, loudly and slowly. One must also disarm the opposing argument; explain away its force while showing it due respect."

Nozick sharpens the "pull" of the two arguments as follows. Suppose the experiment had been done many times before. In every case the Being predicted correctly. Those who took both boxes always got only $1,000; those who took only B2 got $1 million. You have no reason to suppose your case will be different. If a friend were observing the scene, it would be completely rational for him to bet, giving high odds, that if you take both boxes you will get only $1,000. Indeed, if there is a time delay after your choice of both boxes, you know it would be rational for you yourself to bet, offering high odds, that you will get only $1,000. Knowing this, would you not be a fool to take both boxes?

Alas, the other argument makes you out to be just as big a fool if you do not. Assume that B1 is transparent. You see the $1,000 inside. You cannot see into B2, but the far side is transparent and your friend is sitting opposite. He knows whether the box is empty or contains $1 million. Although he says nothing, you realize that, whatever the state of B2 is, he wants you to take both boxes. He wants you to because, regardless of the state of B2, you are sure to come out ahead by $1,000. Why not take advantage of the fact that the Being played first and cannot alter his move?

Nozick, an expert on decision theory, approaches the paradox by considering analogous game situations in which, as here, there is a conflict between two respected principles of choice: the "expected-utility principle" and the "dominance principle." To see how the principles apply, consider the payoff matrix for Newcomb's game [see Figure 93]. The argument for taking only B2 derives from the principle that you should choose so as to maximize the expected utility (value to you) of the outcome. Game theory calculates the expected utility of each action by multiplying each of its mutually exclusive outcomes by the probability of the outcome, given the action. We have assumed that the Being

		BEING	
		MOVE 1 (PREDICTS YOU TAKE ONLY BOX 2)	MOVE 2 (PREDICTS YOU TAKE BOTH BOXES)
YOU	MOVE 1 (TAKE ONLY BOX 2)	$1,000,000	$0
	MOVE 2 (TAKE BOTH BOXES)	$1,001,000	$1,000

Figure 93 Payoff matrix for Newcomb's paradox

predicts with near certainty, but let us be conservative and make the probability a mere .9. The expected utility of taking both boxes is

$$(.1 \times \$1{,}001{,}000) + (.9 \times \$1{,}000) = \$101{,}000.$$

The expected utility of taking only B2 is

$$(.9 \times \$1{,}000{,}000) + (.1 \times \$0) = \$900{,}000.$$

Guided by this principle, your best strategy is to take only the second box.

The dominance principle, however, is just as intuitively sound. Suppose the world divided into *n* different states. For each state *k* mutually exclusive actions are open to you. If in at least one state you are better off choosing *a*, and in all other states either *a* is the best choice or the choices are equal, then the dominance principle asserts that you should choose *a*. Look again at the payoff matrix. The states are the outcomes of the Being's two moves. Taking both boxes is strongly dominant. For each state it gives you $1,000 more than you would get by taking only the second box.

That is as far as we can go into Nozick's analysis, but interested readers should look it up for its mind-boggling conflict situations related to Newcomb's problem. Nozick finally arrives at the following tentative conclusions:

If you believe in absolute determinism and that the Being has in truth predicted your behavior with unswerving accuracy, you should "choose" (whatever that can mean!) to take only B2. For example, suppose the Being is God and you are a devout Calvinist, convinced that God knows every detail of your future. Or assume that the Being has a time-traveling device he can launch into the future and bring back with a motion picture of what you did on that future occasion when you made your choice. Believing that, you should take only B2, firmly persuaded that your feeling of having made a genuine choice is sheer illusion.

Nozick reminds us, however, that Newcomb's paradox does *not* assume that the Being has perfect predictive power. If you believe that you possess a tiny bit of free will (or alternatively that the Being is sometimes wrong, say once in every 20 billion cases), then this may be one of the times the Being has erred. Your wisest decision is to take both boxes.

Nozick is not happy with this conclusion. "Could the difference between one in n and none in n, for arbitrarily large finite n, make this difference? And how exactly does the fact that the predictor is certain to have been correct dissolve the force of the dominance argument?" Both questions are left unanswered. Nozick hopes that publishing the problem "may call forth a solution which will enable me to stop returning, periodically, to it."

One such solution, "to restore [Nozick's] peace of mind," was attempted by Maya Bar-Hillel and Avishai Margalit, of Hebrew University in Jerusalem, in their paper "Newcomb's Paradox Revisited" (see the bibliography for the next chapter). They adopt the same game-theory approach taken by Nozick, but they come to an opposite conclusion. Even though the Being is not a perfect predictor, they recommend taking only the second box. You must, they argue, resign yourself to the fact that your best strategy is to behave *as if* the Being has made a correct prediction, even though you know there is a slight chance he has erred. You know he has played before you, but you cannot do better than to play as if he is going to play after you. "For you cannot outwit the Being except by knowing what he predicted, but you cannot know, or even meaningfully guess, at what he predicted before actually making your final choice."

It may seem to you, Bar-Hillel and Margalit write, that backward causality is operating—that somehow your choice makes the $1 million more likely to be in the second box—but this is pure flim-flam. You choose only B2 "because it is inductively known to correlate remarkably with the existence of this sum in the box, and though we do not assume a causal relationship, there is no better alternative strategy than to behave as if the relationship was, in fact, causal."

For those who argue for taking only B2 on the grounds that causality is independent of the direction of time—that your decision actually "causes" the second box to be either empty or filled with $1 million—Newcomb proposed the following variant of his paradox. Both boxes are transparent. B1 contains the usual $1,000. B2 contains a piece of paper with a fairly large integer written on it. You do not know whether the number is prime or composite. If it proves to be prime (you must not test it, of course, until after you have made your choice), then you get $1 million. The Being has chosen a prime number if he predicts you will take only B2 but has picked a composite number if he predicts you will take both boxes.

Obviously you cannot by an act of will make the large number change from prime to composite, or vice versa. The nature of the number is fixed for eternity. So why not take both boxes? If it is prime, you get $1,001,000. If it is not, you get at least $1,000. (Instead of a number, B2 could contain any statement of a decidable mathematical fact that you do not investigate until after your choice.)

It is easy to think of other variations. For example, there are 100 little boxes, each holding a $10 bill. If the Being expects you to take all of them, he has put nothing else in them. But if he expects you to take only one box — perhaps you pick it at random — he has added to that box a large diamond. There have been thousands of previous tests, half of them involving you as a player. Each time, with possibly a few exceptions, the player who took a single box got the diamond, and the player who took all the boxes got only the money. Acting pragmatically, on the basis of past experience, you should take only one box. But then how can you refute the logic of the argument that says you have everything to gain and nothing to lose if the next time you play you take all the boxes?

These variants add nothing essentially new. With reference to the original version, Nozick halfheartedly recommends taking both boxes. Bar-Hillel and Margalit strongly urge you to "join the millionaire's club" by taking only B2. That is also the view of Kruskal and Newcomb. But has either side really done more than just repeat its case "loudly and slowly"? Can it be that Newcomb's paradox validates free will by invalidating the possibility, in principle, of a predictor capable of guessing a person's choice between two equally rational actions with better than 50 percent accuracy?

ADDENDUM

So many letters poured in about Newcomb's paradox that I asked Robert Nozick if he would be willing to look them over and write a guest column about them. To my delight, he agreed. I packed a large carton with the correspondence and took it along on a visit to the Artificial Intelligence Laboratory at M.I.T. During this visit I had the pleasure of lunching with Nozick on the Harvard Yard and depositing my carton of letters on his desk. Although his column did not run in *Scientific American* until eight months after my column on the topic, it seems appropriate to place it directly after this chapter. It will have a longer addendum and a bibliography.

In 1974, shortly after Nozick's column appeared, Basic Books issued his controversial defense of political libertarianism, *Anarchy, State, and Utopia.* It won the 1975 National Book Award, catapulting Nozick into the ranks of major U.S. philosophers. *Reading Nozick,* a collection of papers attacking and defending him, edited by Jeffrey Paul, came out in 1981. That same year Harvard University Press published Nozick's massive *Philosophical Explanations,* boosting his reputation still higher. It won Phi Beta Kappa's Ralph Waldo Emerson award for the year. As far as I know, Nozick has not written about Newcomb's paradox since he wrote the chapter you are about to read.

CHAPTER FOURTEEN

Reflections on Newcomb's Paradox

By Robert Nozick

Newcomb's problem involves a Being who has the ability to predict the choices you will make. You have enormous confidence in the Being's predictive ability. He has already correctly predicted your choices in many other situations and the choices of many other people in the situation to be described. We may imagine that the Being is a graduate student from another planet, checking a theory of terrestrial psychology, who first takes measurements of the state of our brains before making his predictions. (Or we may imagine that the Being is God.) There are two boxes. Box 1 contains $1,000. Box 2 contains either $1 million or no money.

You have a choice between two actions: taking what is in both boxes or taking only what is in the second box. If the Being predicts you will take what is in both boxes, he does not put the $1 million in the second box. If he predicts you will take only what is in the second box, he puts the $1 million in the second box. (If he predicts you will base your choice on some random event, he does not put the money in the second box.) You know these facts, he knows you know them and so on. The Being makes his prediction of your choice, puts the $1 million in the second box or not and then you choose. What do you do?

There are plausible arguments for reaching two different decisions:

1. *The expected-utility argument.* If you take what is in both boxes, the Being almost certainly will have predicted this and will not have put the $1 million in

	HE PREDICTS YOUR CHOICE CORRECTLY	HE PREDICTS YOUR CHOICE INCORRECTLY
TAKE BOTH	$1,000	$1,001,000
TAKE ONLY SECOND	$1,000,000	$0

Figure 94　Payoff matrix for expected-utility argument

the second box. Almost certainly you will get only $1,000. If you take only what is in the second box, the Being almost certainly will have predicted this and put the money there. Almost certainly you will get $1 million. Therefore (on plausible assumptions about the utility of the money for you) you should take only what is in the second box *[see Figure 94]*.

2.　*The dominance argument.* The Being has already made his prediction and has either put the $1 million in the second box or has not. The money is either sitting in the second box or it is not. The situation, whichever it is, is fixed and determined. If the Being put the $1 million in the second box, you will get $1,001,000 if you take both boxes and $1 million if you take only the second box. If the Being did not put the money in the second box, you will get $1,000 if you take both boxes and no money if you take only the second box. In either case you will do better by $1,000 if you take what is in both boxes rather than only what is in the second box *[see Figure 95]*.

Each argument is powerful. The problem is to epxlain why one is defective. Of the first 148 letters to *Scientific American* from readers who tried to resolve the paradox, a large majority accepted the problem as being meaningful and favored one of the two alternatives. Eighty-nine believed one should take only what is in the second box, 37 believed one should take what is in both boxes — a proportion of about 2.5 to one. Five people recommended cheating in one way or another, 13 believed the problem's conditions to be impossible or inconsistent and four maintained that the predictor cannot exist because the assumption that he does leads to a logical contradiction.

Those who favored taking only the second box tried in various ways to undercut the force of the dominance argument. Many pointed out that if you thought of that argument and were convinced by it, the predictor would (almost certainly) have predicted it and you would end up with only $1,000. They interpreted the dominance argument as an attempt to outwit the predictor. This position makes things too simple. The proponent of the dominance argument does believe he will end up with only $1,000, yet nevertheless he thinks it is best to take both boxes. Several proponents of the dominance principle bemoaned the fact that rational individuals would do worse than irrational ones, but that did not sway them.

	HE PUT $1,000,000 INTO BOX 2	HE DID NOT PUT $1,000,000 INTO BOX 2
TAKE BOTH	$1,001,000	$1,000
TAKE ONLY SECOND	$1,000,000	$0

Figure 95   Payoff matrix for dominance argument

Stephen E. Weiss of Morgantown, W. Va., tried to reconcile the two views. He suggested that following the expected-utility argument maximizes expectation, whereas following the dominance argument maximizes correct decision. Unfortunately that leaves unexplained why the correct decision is not the one that maximizes expectation.

The assumptions underlying the dominance argument — that the $1 million is already in the second box or it is not and that the situation is fixed and determined — were questioned by Mohan S. Kalelkar, a physicist at the Nevis Laboratories of Columbia University, who wrote: "Perhaps it is false to say that the Being has definitely made one choice or the other, just as it is false to say that the electron [in the two-slit experiment] went through one slit or the other. Perhaps we can only say that there is some amplitude that B2 [second box] has $1 million and some other amplitude that it is empty. These amplitudes interfere unless and until we make our move and open up the box. . . . To assert that 'either B2 contains $1 million or else it is empty' is an intuitive argument for which there is no evidence unless we open the box. Admittedly the intuitive evidence is strong, but as in the case of the double-slit electron diffraction our intuition can sometimes prove to be wrong."

Kalelkar's argument makes a version of the problem, in which the second box is transparent on the other side and someone has been staring into it for a week before we make our choice, a significantly different decision problem. It seems not to be. Erwin Schrödinger, in a famous thought experiment, imagined a cat left alone in a closed room with a vial of cyanide that breaks if a radioactive atom in a detector decays. Must a disciple of Niels Bohr assert that the cat is neither alive nor dead, Schrödinger asked, until measurements have been made to decide the case? Even if one accepts the Bohr interpretation of quantum mechanics, however, what choice *does* one make, in Newcomb's problem, when one knows that others can see into the box from the other side and observe whether it is filled or empty?

Many who wrote asserted that the dominance argument assumes the states to be probabilistically independent of the actions and pointed out that this is not true for the two states "The $1 million is in Box 2" and "The $1 million is not in

Box 2." The states would be probabilistically independent of the actions (let us assume) in the matrix for the utility argument, which has the states "He predicts correctly" and "He predicts incorrectly." Here, however, there is no longer dominance. Therefore it appears that the force of dominance principles is undercut. "It is legitimate to apply dominance principles if and only if the states are probabilistically independent of the actions. If the states are not probabilistically independent of the actions, then apply the expected-utility principle, using as the probability-weights the conditional probabilities of the states given in the actions." The quotation is from my original 1970 essay, which formulated this position, then went on to reject it as unsatisfactory for the following reasons.

Suppose a person knows that either man S or man T is his father, but he does not know which. S died of some very painful inherited disease that strikes in one's middle thirties, and T did not. The disease is genetically dominant. S carried only the dominant gene. T did not have the gene. If S is his father, the person will die of the dread disease. If T is his father, he will not. Furthermore, suppose there is a well-confirmed theory that states that a person who inherits this gene will also inherit a tendency toward behavior that is characteristic of intellectuals and scholars. S had this tendency. Neither T nor the person's mother had such a tendency. The person is now deciding whether to go to graduate school or to become a professional baseball player. He prefers (although not enormously) the life of an academic to that of a professional athlete. Regardless of whether or not he will die in his middle thirties, he would be happier as an academic. The choice of the academic life would thus appear to be his best choice.

Now suppose he reasons that if he decides to be an academic, the decision will show that he has such a tendency, and therefore it will be likely that he carries the gene for the disease and so will die in his middle thirties, whereas if he chooses to become a baseball player, it will be likely that T is his father; therefore he is not likely to die of the disease. Since he very much prefers not dying of the disease (as a baseball player) to dying early from the disease (as an academic), he decides to pursue the career of an athlete. Surely everyone would agree that this reasoning is perfectly wild. It is true that the conditional probabilities of the states "S is his father" and "T is his father" are not independent of the actions "becoming an academic" and "becoming a professional athlete." If he does the first, it is very likely that S is his father and that he will die of the disease; if he does the second, it is very likely that T is his father and therefore unlikely that he will die of the disease. But who his father is cannot be changed. It is fixed and determined and has been for a long time. His choice of how to act legitimately affects our (and his) estimate of the probabilities of the two states, but which state obtains (which person is his father) does not depend on his action at all. By becoming a professional baseball player he is not making it less likely that S is his father; therefore he is not making it less likely that he will die of the disease.

This case, and others more clearly including a self-reference that this case may seem to lack, led me to think probabilistic nonindependence was not sufficient to reject the dominance principle. It depends on whether the actions influence or affect the states; it is not enough merely that they affect our judgments about whether the states obtain. How do those who reject the dominance principle for Newcomb's problem distinguish it from those other cases where dominance principles obviously apply even though there is probabilistic nonindependence?

But one must move carefully here. One cannot force a decision in a difficult case merely by finding another similar case where the decision is clear, then challenging someone to show why the decision should be different in the two cases. There is always the possibility that whatever makes one case difficult and the other clear will also make a difference as to how they should be decided. The person who produces the parallel example must not only issue his challenge; he must also offer an explanation of why the difficult case is less clear, an explanation that does not involve any reason why the cases might diverge in how they should be decided. Interested readers can find my additional parallel examples, where dominance is appropriate, plus an attempt to explain why Newcomb's case, although less clear, is still subject to dominance principles in my original essay, "Newcomb's Problem and Two Principles of Choice" in *Essays in Honor of Carl G. Hempel,* (see bibliography).

This obligation to explain differences in the clarity of parallel examples in order to show that no different decision should be made also rests on those who argued in their letters for taking only what is in the second box. For example, it rests on Robert Heppe of Fairfax, Va., who said that the situation "is isomorphic with one in which the human moves first and openly," and on A. S. Gilbert of the National Research Council of Canada, who called the Newcomb case "effectually the same as" one where you act first and an observer attempts to communicate with a "mindreader" in the next room who then guesses your choice, using a payoff matrix identical with Newcomb's.

A large number of those who recommended taking only the second box performed the expected-value calculation and concluded that, provided the probability that the Being was correct was at least .5005, they would take only the second box. Not only did they see no problem at all, but they either maximized expected monetary value or made utility linear with money in the range of the problem. Otherwise the cutoff probability would be different.

William H. Riker of the department of political science at the University of Rochester suggested that people making different decisions merely differed in their utility curves for money. Such persons, however, need not differ in their choices among probability mixtures of monetary amounts in the standard situations in order to calibrate their utilities.

Those who favored taking both boxes made almost no attempt to diagnose the mistakes of the others. An exception is William Bamberger, an economist at Wayne State University. He wrote that the proponent of choosing only the second box "computes not the alternative payoffs of choosing one or two boxes for a given individual, but the average payoff of those who choose two as opposed to the average observed payoff of those who choose one." The problem, of course, is how to compute the probability for a given individual of his payoff for each choice. Should one use the differing conditional probabilities, or ignore them because dominance applies only when the states are probabilistically independent of the action (and so when for each state its conditional probabilities on each act is the same), or ignore them because the conditional probabilities of the state on the acts are to be used only when they represent some process of the act's influencing or affecting which state obtains?

A number of respondents said their choice would depend on whether the predictor made his prediction after they had at least started to consider the problem. If so, they would do their best to decide to take only the second box (so that this data would be available to the predictor), and some added that they hoped they would change their mind at the last minute and take both boxes. (They gave the predictor too little credit.) On the other hand, if the predictor made his prediction before they even considered the problem, these writers believed they would take both boxes, since there was no possibility of their deliberations affecting the prediction that had been made.

Several respondents maintained that if the conditions of the problem could be realized, we might be forced to revise our views about the impossibility of backward causality. Newcomb himself seems to think that special difficulties arise for proponents of backward causality if the predictor writes some term designating an integer on a slip of paper in the second box, with the understanding that you get $1 million only if that integer is a prime. Of course, the predictor writes a prime if, and only if, he predicts that you will take the second box. How can your choice determine whether a number is prime or composite? The advocate of backward causality need not think it does. What your choice affects, in his view, is what term the predictor writes down (or wrote down earlier), not whether the integer it designates is prime or composite.

The reasoning of some of the letters indicates it would be useful to specify precisely the conditions whereby we could discover in which time-direction causality operates. Might one even say that some conditions universally preceding certain decisions are part of the effects of the decision (by backward causality) rather than part of the cause?

Not everyone was willing to choose one or the other action. Among the five respondents who suggested some form of cheating, Robert B. Pitkin, editor of *American Legion Magazine,* speculated that Dr. Matrix, the numerologist, would walk in with a device to scan the contents of the boxes, take the boxes with the money in them and never open an empty box. "He quite naturally succeeded in getting all the money, for the rule of bridge that one peek is worth two finesses applies here too. . . . By introducing a choice which the Being has not anticipated, and is not permitted to take into account, he achieves a stunning victory for free will." (What prevents the Being from taking this into account?)

Other letter writers also struck blows for free will. Nathan Whiting of New York would take both boxes but would open only the first one, leaving the second box unopened. Ralph D. Goodrich, Jr., of Castle Rock, Colo., would take only the first box. Richard B. Miles of Los Altos, Calif., also recommended a "creative" solution: Turn to another person before you make your choice and offer to sell him for $10,000 the contents of whatever box or boxes you choose.

Isaac Asimov wrote: "I would, without hesitation, take both boxes. . . . I am myself a determinist but it is perfectly clear to me that any human being worthy of being considered a human being (including most certainly myself) would prefer free will, if such a thing could exist. . . . Now, then, suppose you take both boxes and it turns out (as it almost certainly will) that God has foreseen this and placed nothing in the second box. You will then, at least, have expressed your willingness to gamble on his nonomniscience and on your own free will and will have willingly given up a million dollars for the sake of that willingness — itself a snap of the finger in the face of the Almighty and a vote, however futile, for free will. . . . And, of course, if God has muffed and left a million dollars in the box, then not only will you have gained that million but *far more important* you will have demonstrated God's nonomniscience. If you take only the second box, however, you get your damned million and not only are you a slave but also you have demonstrated your willingness to be a slave for that million and you are not someone I recognize as human." (No one wrote to argue for taking only the second box on the grounds that either it results in getting $1 million or it demonstrates the Being's fallibility, either of which is desirable.)

Those who held that the conditions of the problem could not be realized were of two types. There were those who believed the situation to be physically impossible because the Being could not predict all the information input of every light signal that would arrive at your eyes in the appropriate time interval.

("To gain such knowledge the Being must have a physical agency for collecting information that travels faster than the speed of light," wrote George Fredericks, a physicist at the University of Texas.) And there were those who argued that if the room is closed, the problem reduces to that of Maxwell's demon — a suggestion made by Fredericks and by John A. Ball of the Harvard College Observatory.

Those who believed the conditions of the problem to be inconsistent as well as physically impossible said that the almost certain predictability of decisions was inconsistent with free will, and therefore with making choices, yet the problem assumed that genuine choices could be made. This is a hard argument to drive through because it appears to be the choices that are predicted. The relevant connections are difficult to get straight. Predictability of decisions does not logically imply determinism under which the decisions are caused (for example, the possibility of backward causality where an uncaused decision causes an earlier prediction, or "seeing ahead" in time in a block universe).

Nor, we should note in passing, does determinism entail predictability, even in principle. Events could be fixed in accordance with scientific laws that are not recursive. Is determinism incompatible with free will? It seems to many to be so, yet the argument that determinism is incompatible with responsibility for action, which free will implies, depends on a notion of responsibility insufficiently worked out to show precisely how the connections go. Some say merely that a free act is an uncaused one. Yet being uncaused obviously is not sufficient for an act to be free; one surely would not be responsible for such an action. What other conditions, then, must be satisfied by an uncaused act if it is to be a free one? The literature on free will lacks a satisfactory specification of what a free action would be like (given that "uncaused" is not enough). Perhaps if we were given this specification of additional conditions, they would turn out to be sufficient apart from the action's being uncaused.

Another problem will help to exhibit some complicated relations between free will and determinism. It has been asserted (by C. S. Lewis, for instance) that no determinist rationally can believe in determinism, for if determinism is true, his beliefs were caused, including his belief in determinism. The idea seems to be that the causes of belief, perhaps chemical happenings in the brain, might be unconnected with any reasons for thinking determinism true. They might be, but they need not be. The causes might "go through" reasons and be effective only to the extent that they are good reasons. In the same way it might be a causal truth about someone that he is convinced only by arguments that constitute specified types of good reasons (deductive, inductive, explanatory and so on).

Some philosophers have argued recently that we know some statement p only if part of the cause (or more broadly the explanation) of our believing p is, if we pursue the story far enough, the fact that p is true. You know this page is before you now only if its being there is part of the explanation of why you believe it is there. If psychologists are stimulating your brain to create the illusion that you are seeing a printed page, you would not really know there is a page before you even if a psychologist happened to have left one on the table in front of you. The page's being there would not play the proper causal role in the story of your belief. If we do not mind our beliefs being caused by the facts, and indeed find it somewhat plausible to think we have knowledge only to the extent that they are, then we may also find it less disturbing that our actions are caused by certain types of facts holding in the world — for example, the fact that it would be better to do one thing rather than another. To say this, of course, is not to present a theory of free action; it is merely to hint that it may be possible to remove the sting of determinism. This approach is a comfortable one when we act correctly, but it is difficult to see how it can be extended plausibly to wrong acts where questions of responsibility are particularly pressing.

Proponents of the C. S. Lewis position might reply that the determinist should not feel so comfortable. Even though he says he is caused to believe in determinism (and anything else) by what are good reasons, he must also maintain that he is caused to believe that such reasons are good reasons. He may have a second set of reasons for believing the first set of reasons are good. Now, however, his opponent can raise the same question as before. Why does he believe the second set of reasons? The determinist must end either by finding self-supporting reasons (which say of themselves that they are good reasons) or by admitting that the best explanation of why he believes they are good reasons is that they are. This surely leaves his opponent unsatisfied, and the match seems to be a draw.

Those who believe in free will find themselves in similar dilemmas. Kurt Rosenwald of Washington wrote: "When I was 19 or 20, I thought about the free-will problem . . . and I came to this conclusion: If we make an exhaustive study of that problem, and finally arrive at the result that our will is free, we still will not know whether our will is indeed free or our mind is of such a nature that we have to find our will to be free, although it is not, in fact, free. This became one of my reasons for studying not philosophy but the natural sciences. Thinking about it now, 50+ years later, it still seems to me that I was right." But does not the possibility that we are caused to believe in false conclusions apply also to conclusions in the natural sciences? And to the verdict of 50+ years later?

I published my original essay after thinking about Newcomb's problem intermittently for five years. In that essay I expressed the hope that someone would come forth with a solution to the problem that would enable me to stop returning to it. It is not surprising that no one did, yet it is surprising (to me) that the mere act of publishing Newcomb's problem, and sending my thoughts on it into the world, rid me of it. That is, I was rid of it until the problem was presented in *Scientific American* and I was invited to read more than 650 pages of letters about it. Unfortunately the letters do not, in my opinion, lay the problem to rest. And they have started me thinking about it again! You can't win.

ADDENDUM (BY M. G.)

Although the growing literature on Newcomb's problem proves that philosophers are still far from agreement on how to handle it, let me set down some tentative personal views.

My sympathies are with those who say the predictor cannot exist. Even if strict determinism in some sense holds for every event in the history of the universe, I believe that certain events are in principle unpredictable when predictions are allowed to interact causally with the event being predicted. We have here, I am persuaded, something analogous to the resolution of semantic paradoxes. Contradictions arise whenever a language is allowed to talk about the truth or falsity of its own statements, or when sets are allowed to be members of themselves. We can escape the semantic paradoxes by permitting talk about the truth of a sentence only in a metalanguage. "This sentence is false" simply is not a sentence. The notorious paradox of the barber who shaves every person and only those persons who do not shave themselves, and who himself belongs to the set of persons, is a barber who cannot exist. It is not logically inconsistent to suppose that the future is totally determined, whether or not an omniscient God exists, but as soon as we permit a superbeing to make predictions that interact with the event being predicted, we encounter contradictions that render the existence of such a superpredictor impossible.

Consider the simplest case. A superbeing knows that when you go to bed next Thursday you will take off your shoes. If the superbeing keeps this knowledge from you, there is no problem; but if the superbeing informs you of the prediction, you can falsify it easily by going to bed with your shoes on. At this point we touch the mystery of free will, about which I have a chapter in my *Whys of a Philosophical Scrivener* (Morrow, 1983). I agree with those who say that Newcomb's problem in no way settles the question of whether the future is completely determined, but I do maintain that it brings us face to face with the eternal, and to me unanswerable, problem of defining what is meant by free choice.

Although I don't believe it, the state of the world a hundred years from now may be determined in every detail by the state of the world now. Innumerable future events obviously can be predicted with almost certain accuracy, but other events are the outcome of such complex causes that even if determinism is true it seems likely there is no possible way they could be predicted by any technique faster than allowing the universe itself to unroll to see what happens. (We leave aside the notion of a God outside of time who sees the past and future simultaneously, whatever that means.) All this is by the way. The main point is that when a prediction interacts with the predicted event, whether human wills are involved or not, logical contradictions can arise. A familiar example is the supercomputer asked to predict if a certain event will occur in the next three minutes. If the prediction is no, it turns on a green light. If yes, it turns on a red light. The computer is now asked to predict whether the green light will go on. By making the event part of the prediction, the computer is rendered logically impotent.

It is my view that Newcomb's predictor, even if accurate only 51 percent of the time, forces a logical contradiction that makes such a predictor, like Bertrand Russell's barber, impossible. We can avoid contradictions arising from two different "shoulds" (should you take one or two boxes?) by stating the contradiction as follows. One flawless argument implies that the best way to maximize your reward is to take only the closed box. Another flawless argument implies that the best way to maximize your reward is to take both boxes. Because the two conclusions are contradictory, the predictor cannot exist. Faced with a Newcomb decision, I would share the suspicions of Max Black and others that I was either the victim of a hoax or of a badly controlled experiment that had yielded false data about the predictor's accuracy. On this assumption, I would take both boxes.

But, you may ask, how would I decide if I made what I would regard as a counterfactual posit that the predictor was what it was claimed to be? I suppose if I could persuade myself that the predictor existed I might take only the closed box even though it would be logically irrational. But I cannot so persuade myself. It is as if someone asked me to put 91 eggs in 13 boxes, so each box held seven eggs, and then added that an experiment had proved that 91 is prime. On that assumption, one or more eggs would be left over. I would be given a million dollars for each leftover egg, and 10 cents if there were none. Unable to believe that 91 is a prime, I would proceed to put seven eggs in each box, take my 10 cents and not worry about having made a bad decision.

BIBLIOGRAPHY

"Newcomb's Problem and Two Principles of Choice." Robert Nozick in *Essays in Honor of Carl G. Hempel,* edited by Nicholas Rescher. Humanities Press, 1969.

Paradoxes of Rationality: Theory of Metagames and Political Behavior. Nigel Howard. MIT Press, 1971, pages 168–184.

"Newcomb's Paradox Revisited." Maya Bar-Hillel and Avishai Margalit in *The British Journal for the Philosophy of Science*, Vol. 23, 1972, pages 295–304. Defends the pragmatic decision. "We hope to convince the reader to take just the one covered box, and join the millionaire's club!"

"The Unpredictability of Free Choices." George Schlesinger in *The British Journal for the Philosophy of Science*, Vol. 25, 1974, pages 209–221. Argues that no amount of inductive evidence can support the belief that the predictor is capable of better than chance in its predictions, and that the paradox brings out the fundamental unpredictability of human choices and the reality of free will. It is best to take both boxes.

"Newcomb's Many Problems." Isaac Levi in *Theory and Decision*, Vol. 6, 1975, pages 161–175.

"Newcomb's Paradox." James Cargile in *The British Journal for the Philosophy of Science*, Vol. 26, 1975, pages 234–239. Contends that there is "no basis for determining one course of action as the right one without additional information."

"Newcomb's Problem and the Prisoners' Dilemma." Steven J. Brams in *The Journal of Conflict Resolution*, Vol. 19, 1975, pages 596–619. Argues the close relationship of Newcomb's problem to the prisoner's dilemma. Applying metagame theory, Brams maintains that if you believe empirical evidence indicates the predictor's accuracy is greater than .5005, you should take only one box; if less than .5005, you should be indifferent as to your choice.

"A Paradox of Prediction." Steven J. Brams in *Paradoxes in Politics*, Chapter 8. Free Press, 1976. Enlarges on the previous entry.

"Newcomb's 'Paradox.' " T. M. Benditt and David J. Ross in *The British Journal for the Philosophy of Science*, Vol. 27, 1976, pages 161–164. Attack's Schlesinger's reasoning and concludes that "there is only one rational choice . . . namely choose box II alone."

"Unpredictability: A Reply to Cargile and to Benditt and Ross." George Schlesinger in *The British Journal of the Philosophy of Science*, Vol. 27, 1976, pages 267–274. The author defends his earlier position.

"How to Make a Newcomb Choice." Don Locke in *Analysis*, January, 1978, pages 17–23. Takes the view that the situation is not sharply enough defined to make decision theory applicable. It is best to take both boxes,

> "confident that whether Box One is full or empty I cannot suffer from that choice, hopeful that I might at least demonstrate freedom of choice in the presence of a Newcomb Predictor, but resigned to receive a mere thousand where less rational mortals, who from the evidence of *Scientific American*

(March, 1974, page 102) outnumber the rational by the order of 5 to 2, stand to gain a million. Perhaps it is precisely because I am a rational man that the Predictor is able to predict my choice, and leave Box One empty. Perhaps, in this respect at least, I would be better off were I less rational than I am. The penalties of philosophy are no less than its pleasures."

"Counterfactuals and Two Kinds of Expected Utility." Allan Gibbard and William Harper in *Foundations and Applications of Decision Theory,* edited by A. A. Hooker and others. Reidel, 1978. Reprinted in *Ifs,* edited by William Harper and others, Reidel, 1981.

"Prisoners' Dilemma Is a Newcomb Problem." David Lewis in *Philosophy and Public Affairs,* Vol. 8, 1979, pages 235–240.

"Newcomb's Paradox and the Principle of Maximizing Conditional Expected Utility." Ellery Eells. Ph.D. dissertation, University of California, Berkeley, June, 1980.

"Counterfactuals and Newcomb's Problem." T. Horgan in *The Journal of Philosophy,* Vol. 78, 1981, pages 331–356.

"Newcomb's Paradox." Fred Alan Wolf in *Taking the Quantum Leap.* Harper and Row, 1981. The funniest attempt at resolution yet, by a physicist deep into the paranormal and a defender of the solipsistic view that the world "out there" really isn't there until it is created by your observation. Since the money is not in the closed box until you perceive it, you have to put it there by choosing and opening the box. "It is your act of observation that resolves the paradox. Choosing both boxes creates box *R* empty. Choosing box *R* creates it one million dollars fuller. . . . your choices create the alternate possibilities as realities."

"A Note on Newcombmania." Isaac Levi in *The Journal of Philosophy,* Vol. 79, 1982, pages 337–342.

Rational Decision and Causality. Ellery Eells. Cambridge University Press, 1982.

"The Wrong Box." Isaac Levi in *The Journal of Philosophy,* Vol. 80, 1983, pages 534–542.

"Newcomb's Problem: Recalculations for the One-Boxer." Roy A. Sorensen in *Theory and Decision,* Vol. 15, 1983, pages 399–404.

"Newcomb's Problem." Steven J. Brams in *Superior Beings.* Springer-Verlag, 1983, pages 46–54. Summarizes his earlier views.

"Newcomb's Problem Demystified." Max Black in *The Prevalence of Humbug and Other Essays.* Cornell University Press, 1983, Chapter 8. Argues that the existence of the predictor is so impossible that anyone faced with a Newcomb decision "would be well advised to suspect fraud, and to play safe by taking both boxes."

"Newcomb's Many Solutions." Ellery Eells in *Theory and Decision,* Vol. 16, 1984, pages 59–105.

"Rationality Revisited." Reed Richter in *The Australasian Journal of Philosophy,* Vol. 62, 1984, pages 392–403.

"The Iterated Versions of Newcomb's Problem and the Prisoner's Dilemma." Roy A. Sorensen in *Synthese,* Vol. 63, 1985, pages 157–166.

"Levi's 'The Wrong Box.' " Ellery Eells in *The Journal of Philosophy,* Vol. 82, 1985, pages 91–106.

Paradoxes of Rationality and Cooperation: Prisoner's Dilemma and Newcomb's Problem. Edited by Richmond Campbell and Lanning Sowden. University of British Columbia Press, 1985.

CHAPTER FIFTEEN

Reverse the Fish and Other Problems

1. THE GUNPORT PROBLEM

Determining the shortest possible game of cram (see the previous chapter) is equivalent to what Bill Sands called the domino "gunport problem." (See his article, "The Gunport Problem," in *Mathematics Magazine,* Vol. 44, 1971, pages 193–196)

The problem is simply stated: What is the maximum number of 1-by-1 "holes" that can be obtained by arranging dominoes on an *m*-by-*n* field? It is assumed that *m* and *n* are each greater than 1.

Sands was able to prove that the number of holes cannot exceed the number of dominoes. He also showed that if either side of the field is a multiple of 3, a repeated pattern provides a simple way of achieving the maximum number of holes *[see Figure 96].* In other words, if one side of the field is of the form 3*k*, the maximum number of holes is *mn*/3. Otherwise the maximum number of holes must be less than this.

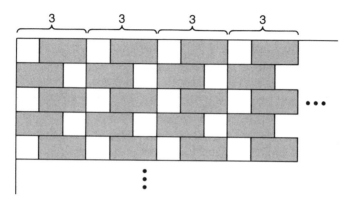

Figure 96   Pattern for maximizing "gunports" when one side of rectangle is 3*k*

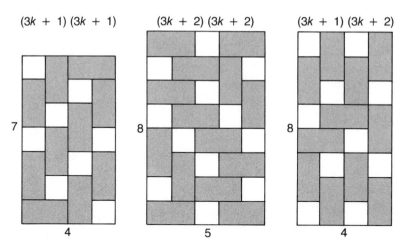

Figure 97 Maximum gunports on rectangles having no sides of form $3k$

Murray Pearce of Bismarck, N.D., writing in the November 1973 issue of the London monthly *Games & Puzzles*, conjectured that if neither side is $3k$ but both are equal modulo 3 (that is, both are either $3k + 1$ or $3k + 2$), the maximum number of holes is $(mn - 4)/3$, and if one side is $3k + 1$ and the other $3k + 2$, the maximum is $(mn - 2)/3$. Examples of how the predicted maximum can be obtained for the three types of fields that have no side equal to $3k$ are shown in Figure 97.

Pearce's formulas set a maximum of 26 holes for the 8-by-10 field. Sands confessed he was unable to do better than 24 holes, using 28 dominoes. Can the reader find a 26-hole solution, using 27 dominoes?

2. FIGURES NEVER LIE

An old burlesque routine involves two simpleminded men who divide 28 by 13 to get 7, then verify this result by multiplying 13 by 7 to get 28 and finally double-check it by adding 13 seven times to get 28. This is how Irvin S. Cobb told the story in his anthology of 366 jokes (one for leap year), *A Laugh a Day Keeps the Doctor Away* (1923):

"Three patricians of the coal yards fared forth on mercy bent, each in his great black chariot. Their overlord, the yard superintendent, had bade them deliver to seven families a total of twenty-eight tons of coal equally divided.

"Well out of the yards, each with his first load, Kelly and Burke and Shea paused to discuss the problem of equal distribution — how much coal should each family get?

"'Tis this way,' argued Burke. ''Tis but a bit of mathematics. If there are 7 families an' 28 tons o' coal ye divide 28 by 7, which is done as follows: Seven into 8 is 1, 7 into 21 is 3, which makes 13.' He triumphantly exhibited his figures made with a stubby pencil on a bit of grimy paper:

$$7/28/13$$
$$\underline{7}$$
$$21$$
$$\underline{21}$$
$$00$$

"The figures were impressive but Shea was not wholly convinced. 'There's a easy way o' provin' that,' he declared. 'Ye add 13 seven times,' and he made his column of figures according to his own formula. Then, starting from the bottom of the 3 column, he reached the top with a total of 21 and climbed down the column of 1's, thus; '3, 6, 9, 12, 15, 18, 21, 22, 23, 24, 25, 26, 27, 28.' 'Burke is right,' he announced with finality.

"This was Shea's exhibit:

$$13$$
$$13$$
$$13$$
$$13$$
$$13$$
$$13$$
$$\underline{13}$$
$$28$$

" 'There is still some doubt in me mind,' said Kelly. 'Let me demonstrate in me own way. If ye multiply the 13 by 7 and get 28, then 13 is right.' He produced a bit of stubby pencil and a sheet of paper. ' 'Tis done in this way,' he said. 'Seven times 3 is 21; 7 times 1 is 7, which makes 28. ' 'Tis thus shown that 13 is the right figure and ye're both right. Would ye see the figures?'

"Kelly's feat in mathematics was displayed as follows;

$$13$$
$$\underline{7}$$
$$21$$
$$\underline{7}$$
$$28$$

" 'There is no more argyment,' the three agreed, so they delivered thirteen tons of coal to each family."

The comedian Flournoy Miller made effective use of the routine and published his version of it in his book *Shufflin' Along*. A few years ago Flip Wilson did the bit on his television show and was sued by Miller's daughter for unauthorized use of the material. The case was apparently settled out of court.

"Is there something special about the numbers 7, 13 and 28?" asked the late William R. Ransom, a mathematician at Tufts University. The answer is no. There are just 22 triplets of numbers — one number is a single digit, the other two are two digits each — that can be substituted for 7, 13 and 28 without changing a single word in the routine. Readers are asked to list the 22 triplets.

3. FUNCTIONAL FIXEDNESS

Past experience sometimes has a negative effect on creative thinking. When this involves a difficulty in seeing how a familiar object can be used in an unorthodox way, psychologists call it a manifestation of "functional fixedness." Here are two problems, familiar to psychologists, that illustrate the concept:

You are seated at a bare table and given six objects: a board, pliers open to maximum extent, two small metal angle irons with screw holes, a peg and a length of wire that has been used to bind the peg firmly to the board *[see Figure 98]*. How can you arrange these objects so that the board becomes a horizontal stand several inches above the tabletop and firm enough to support a vase of flowers?

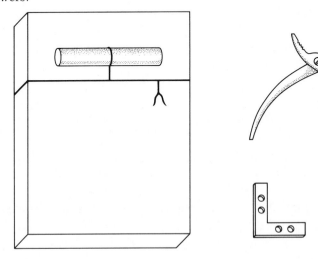

Figure 98   The stand problem

Figure 99 The string problem

You are in a bare room. Two strings hang from the ceiling *[see Figure 99]*. Your problem is to tie the ends together. When you grasp one end of the string, however, the other dangles many feet beyond your reach. You are not allowed to use anything you are wearing or have on your person (such as your stockings for the purpose of swinging them to catch a string), but you may use any or all of three objects on the floor: a table-tennis ball, a small horseshoe magnet and a postage stamp.

4. MONOCHROMATIC CHESS

Here is another brilliant and unorthodox chess problem by Raymond Smullyan. Figure 100 shows the position of an end game with only five men on the board: the black and white kings, two white pawns and one pawn of unknown color *(shown in gray)*. During the course of the game no piece has moved from a square of one color to a square of another color. Is the unknown pawn black or white?

5. THE TWO BOOKCASES

Robert Abes, of the Courant Institute of Mathematical Sciences at New York University, originated this problem. A room 9 by 12 feet contains two bookcases that hold a collection of rare erotica. Bookcase *AB* is 8½ feet long, and bookcase *CD* is 4½ feet long. The bookcases are positioned so that each is centered along its wall and one inch from the wall.

The owner's young nephews are coming for a visit. He wishes to protect them and the books from each other by turning both bookcases around to face the

BLACK

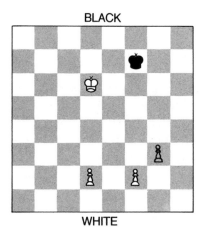

WHITE

Figure 100   Smullyan's monochromatic problem

wall. Each bookcase must end up in its starting position but with its ends reversed *[see Figure 101]*. The bookcases are so heavy that the only way to move them is to keep one end on the floor as a pivot while the other end is swung in a circular arc. The bookcases are narrow from front to back, and for purposes of the problem we idealize them to straight line segments. The ends of the bookcases cannot pass through walls in mid-swing, or through each other. What is the minimum number of swings required to reverse the two bookcases?

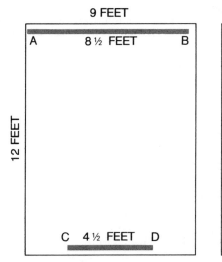

Figure 101   The bookcases problem

6. IRRATIONAL PROBABILITIES

It is very easy to use a penny as a randomizer for deciding between two alternatives with probabilities expressed by rational fractions. Suppose you wish to decide between A and B, with a probability of 3/7 for A and 4/7 for B. The number of equally likely ways a penny can fall when flipped n times is 2^n, so three flips of the coin give eight possible triplets: HHH, HHT, HTH and so on. Eliminate one triplet; then pick any three of the remaining seven and designate them triplets that decide for A. The other four triplets decide for B. Flip the penny three times. If the result is the eliminated triplet, ignore it and flip three more times. Eventually you will flip one of the seven triplets. The chance that this will be a triplet in the set of three is clearly 3/7, with 4/7 as the probability that it will be in the set of four.

The procedure is easily extended to a decision between n alternatives, each with a rational probability. Suppose there are three alternatives with the probabilities A = 1/3, B = 1/2 and C = 1/6. Use the above procedure to decide between 1/3 and 2/3 (the sum of 1/2 and 1/6). If the decision is for A, you are finished. Otherwise you must continue by deciding between B and C. To do so, divide 1/2 (B's fraction) by 2/3 to obtain 3/4, and divide 1/6 (C's fraction) by 2/3 to obtain 1/4. The penny is used as before to decide between B = 3/4 and C = 1/4. The procedure obviously generalizes to n alternatives, provided that the probabilities are rational fractions.

Moreover, the coin need not be a fair one. Suppose it is biased and falls heads with a probability of $1/\pi$. The probability of heads followed by tails remains equal to the probability of tails followed by heads, and so you simply flip doublets, ignoring HH and TT. Let HT count for heads and TH count for tails. With this new definition of heads and tails, each equally likely, the biased coin clearly can be used for deciding between n alternatives, each with rational probabilities.

Suppose, now, you wish to decide between n alternatives, each with an *irrational* probability. For example: A is the fractional part of the square root of 2, B is the fractional part of π and C = 1 − (A + B). If you can decide between two irrational probabilities using a fair coin, you can do it with a biased coin by redefining heads and tails as explained; and if you can decide between two irrational alternatives, you can decide between any number of irrational alternatives by the method given for n rational alternatives.

But how can a coin be used to decide between two irrational probabilities? Let us focus the problem with a precise example. A = .1415926535 . . . , the fractional part of π. B = .8584073464 . . . , which is 1 − A. You wish to decide between *A* and *B* by flipping a fair coin. A delightful procedure for doing this, which applies to all irrational fractions, was recently devised by Persi Diaconis. It will be disclosed in the answer section. (Hint: The method makes use of binary notation.)

7. WHO'S BEHIND THE MAD HATTER?

The following problem, by John F. Collins of Santa Monica, Calif., appeared in the August 1968 issue of *Word Ways*.

"The March Hare and the Mad Hatter were sipping their eggnog and watching the crowd when Alice happened to glance in the Hare's direction and ask, 'Why are you giving me such an angry look?'

" 'I'm not *giving* it to you, I'm giving it *back*,' replied the Hare.

" 'I didn't look crossly at you.'

" 'Well, *somebody* did,' the Hare said, turning to glare at the Hatter.

"Just then, someone came up from behind and put his hands over the Hatter's eyes.

" 'Guess who!', said the newcomer in a thin, flat voice.

"The Hatter froze for a moment and declared, rather coldly, 'I have no use for practical jokers.'

" 'Ha! Neither have I,' retorted the stranger, still keeping his hands over the Hatter's eyes.

"At that, the Hatter seemed to accept the challenge of the game and started asking a series of questions in a manner that mingled hope with care.

"Question: 'Ahem. Would you, by chance, be in a black suit this evening?'

"Answer: 'I would, but not by chance, by design.'

"Q. 'I presume you're a member of all the posh clubs?'

"A. 'Afraid not. Never even been invited.'

"Q. 'Surely you're better than average?'

"A. 'Yes, indeed!'

"Q. 'Not spotted, I hope?'

"A. 'Knock wood.'

"Q. 'Married?'

"A. 'No, happy.' "

Who is behind the Mad Hatter?

8. REVERSE THE FISH

This charming brainteaser for children is well known in Japan but not in this country. I found it in a Japanese puzzle book by kobon Fujimura. Arrange eight

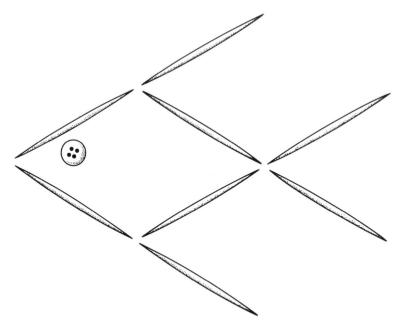

Figure 102   The toothpick puzzle

toothpicks and one button as shown in Figure 102. Now see if you can change the position of just three toothpicks and the button so that the fish looks exactly the same as before except it is now swimming in the opposite direction.

9. THE INTERSECTING CIRCLES

This is one of those elegant theorems in old-fashioned plane geometry that seem at first to be exceedingly difficult to establish but that yield readily to the right insight. Three circles of unit radius, with centers at *X*, *Y*, *Z*, intersect at a common point, *O [see Figure 103]*. The problem is to prove that the other three intersection points, *A*, *B*, *C*, lie on a circle that also has a unit radius. The problem comes by way of Frank R. Bernhart.

ANSWERS

1. Figure 104 shows one way of placing 27 dominoes on an 8-by-10 field to form 26 holes. Found by Capt. John C. Huval, it was published in *Mathematics Magazine* for November 1972. Many trivial variations can be produced by sliding one domino, by switching to adjacent dominoes or by rotating a 3-by-3 pattern of three dominoes.

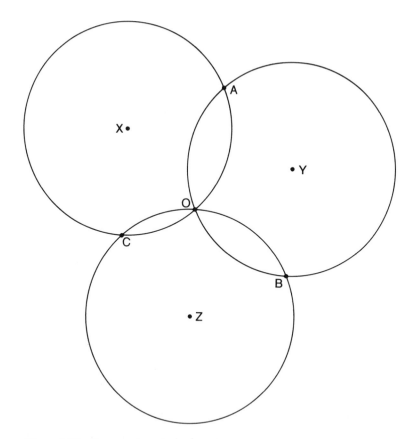

Figure 103 Intersecting-circle theorem

Kenneth M. Brown and Jon Petersen each proved that Murray Pearce's formulas for the gunport problem cannot be exceeded. It remains an open question whether there are rectangles for which Pearce's upper bounds cannot be achieved. Petersen and Douglas W. Oman independently found a procedure showing that all rectangles with areas smaller than 224 could meet the upper bounds. The general case is undecided, with the 14-by-16 rectangle being a likely candidate for the smallest counterexample. According to Pearce's

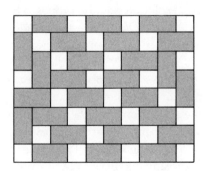

Figure 104 A solution to the gunport problem

conjecture, it should be possible to cover it with 75 dominoes that leave 74 holes.

2. The 22 triplets that can be substituted for 7, 13 and 28 in Irvin S. Cobb's story are

$$12 \div 2 = 15, \quad 15 \div 3 = 14, \quad 16 \div 4 = 13,$$
$$14 \div 2 = 25, \quad 18 \div 3 = 24, \quad 24 \div 4 = 15,$$
$$16 \div 2 = 35, \quad 24 \div 3 = 17, \quad 28 \div 4 = 25,$$
$$18 \div 2 = 45, \quad 27 \div 3 = 27, \quad 36 \div 4 = 18,$$

$$15 \div 5 = 12, \quad 18 \div 6 = 12, \quad 28 \div 7 = 13,$$
$$25 \div 5 = 14, \quad 36 \div 6 = 15, \quad 49 \div 7 = 16,$$
$$35 \div 5 = 16, \quad 48 \div 6 = 17,$$
$$45 \div 5 = 18, \qquad\qquad\qquad 48 \div 8 = 15.$$

Readers interested in how William R. Ransom solved this problem will find it explained in his delightful but little-known book, *One Hundred Mathematical Curiosities* (J. Weston Walch, 1955). Using more liberal interpretations of the dialogue in the old burlesque routine, Joseph H. Engel, Sumner Shapiro and Alan Wayne found other triplets of figures that could be added to the 22 that satisfy a strict interpretation of the dialogue. Wayne recalled having seen the routine performed several times on stage by the Abbott and Costello comedy team.

3. The board is supported by the pliers and wooden peg [see Figure 105]. To tie together the ends of the two hanging cords, tie the magnet to one end and start the cord swinging. Hold the end of the other string and catch the swinging magnet.

For the shelf-making problem E. N. Adams, Bill Kruger and Susan Southall each showed how the pliers could be opened and wired to the peg to make a tripod, and also how the angle irons could be wired flat to each end of the peg to make a stand. The second solution was also proposed by R. C. Dahlquist, P. C. Eastman and Ronald C. Read. Don L. Curtis threaded the wire through end holes of the angle irons and with the pliers tightened the wire around the board so that the angle irons were rigidly perpendicular to the board. He sent a photograph to prove that the stand supported a heavy vase of flowers. Robert Rosenwald and Allan Kiron each thought of using the pliers as a hammer for knocking corners of the angle irons into the board to make supports.

Several readers spotted a mistake in the illustration for the problem of knotting the ends of two hanging strings. The strings are too short to be tied. Paul Nelles considered the situation in which a side wall is so close to each string that if the magnet were tied to one end it woud collide with the wall when the

Figure 105   Supporting the board

string is swung. He suggested tying the table-tennis ball to the cord (or fastening it with the stamp), then swinging the string so that the ball bounces off the wall. Michael McMahon suggested using the stamp to attach the ball to the string and then, with a corner of the magnet, addressing the ball and mailing it to the other side of the room.

4. The key to Raymond Smullyan's monochromatic chess problem lies in the position of the two white pawns. We were told that no piece has moved to a square of a different color; therefore the only way the white king could have escaped from his home square is by castling. The castling must have been on the king's side; otherwise the white rook would have moved from a black square to a white one. If the pawn of unknown color is white, it must have been a rook's pawn that moved to its present square by capturing. But if this was what happened, the white king could not have reached its present position. The rook's pawn, before it made its capture, would have confined the king to KN1 and in its present position would confine the king to KN1 and KR2. Therefore the pawn in question is black.

The black and white sides of the chessboard were properly identified in Raymond Smullyan's monochromatic chess problem, but several readers asked themselves whether the problem could still be solved if the sides were reversed. Two readers independently sent the following "proof" that the uncolored pawn still must be black. Assume that the pawn is white. To reach the position shown in the problem, the pawn must move at least three times: a first move of two squares, then two captures. The other two white pawns must make at least four moves each to reach their positions. This includes six captures. Thus at least eight captures of black pieces, all on black squares, must be made by the three white pawns. At the start of the game Black has eight pieces on black squares. But one of them, the king's knight, cannot move from its original cell. Black therefore has only seven white pieces on black cells that are available for capture. The initial assumption must be false. The uncolored pawn is black.

Unfortunately the proof is false also. William J. Butler, Jr., sent a legal game, conforming to the monochromatic proviso of the problem, in which the sides are reversed and the position shown in the problem is reached with the uncolored pawn being white. The flaw in the above "proof" is that it overlooks the fact that black pawns on initial white cells can be captured *en passant* by white pawns on black cells.

5. Eight swings are enough to reverse those two bookcases. One solution: (1) Swing end *B* clockwise 90 degrees; (2) swing *A* clockwise 30 degrees; (3) swing *B* counterclockwise 60 degrees; (4) swing *A* clockwise 30 degrees; (5) swing *B* clockwise 90 degrees; (6) swing *C* clockwise 60 degrees; (7) swing *D* counterclockwise 300 degrees; (8) swing *C* clockwise 60 degrees.

"If you moved bookcase *AB* in fewer than five swings," writes Robert Abes, who originated this problem, "then you put an end through a wall in midswing, or (more likely) wound up with its front side still facing out. If you moved bookcase *CD* without a 300-degree second swing, you either wasted a swing or scooped a hollow out of a wall. Thanks to Jim Lewis for helping me move the large bookcase."

Wayne E. Russell noted that the bookcase problem did not rule out the possibility that the room was much higher than the cases. He showed that by raising one end of the large case high enough it could be reversed in three swings. Johannes Sack discovered the surprising fact that the minimum-move solution does not correspond to a solution with a minimum expenditure of energy. The given three-move reversal of the small bookcase carries the case at least 33 feet. If four moves are used (*D* counterclockwise 90 degrees, *C* counterclockwise 60 degrees, *D* counterclockwise 60 degrees, *C* clockwise 30 degrees), the case is carried only 18.8 feet, an energy saving of 43 percent.

6. Here is how a coin can be used to decide between alternatives *A* and *B* with probabilities expressed by any rational or irrational fraction.

Notational rules:

a. Express *A* as an endless binary fraction.

b. Number the digits 1, 2, 3, 4, . . . and similarly number the flips of the coin. The *n*th digit is called the "corresponding digit" of the *n*th flip.

c. Let the value of each flip be 1 for heads, 0 for tails.

Procedural rules:

a. If the value of a flip equals its corresponding digit, flip again.

b. If the value of a flip is less than its corresponding digit, stop. This decides for *A*.

c. If the value of a flip is more than its corresponding digit, stop. This decides for *B*.

Let us see how this works when A is 1/3 and B is 2/3. In binary form $A = .01010101$ . . . and $B = .10101010$. . . . The sequence of flips stops with a decision for A if and only if tails (value 0) appears on a flip whose corresponding digit is 1 in the endless binary fraction for A. The 1's are in even positions; therefore the probability of this happening is $1/2^2 + 1/2^4 + 1/2^6 + ...$

The sum of this series is .01010101. . . . This is obvious when we consider its binary fractions:

$$
\begin{aligned}
1/4 &= .01 \\
1/16 &= .0001 \\
1/64 &= .000001
\end{aligned}
$$

$$
\begin{array}{cc}
\cdot & \cdot \\
\cdot & \cdot \\
\cdot & \cdot \\
\hline
& .010101
\end{array}
$$

Similarly, the sequence of flips stops with a decision for B if and only if heads (value 1) appears on a flip corresponding to 0 in the endless fraction for A. The 0's are in odd positions; therefore the probability of this happening is $1/2 + 1/2^3 + 1/2^5 + ...$ This is the same as summing $.1 + .001 + .00001 + ...$, a series that just as obviously adds to .10101 . . . = 2/3.

The specific problem given was to decide between A equaling the fractional part of π and B equaling $1 - A$. First express A as a binary fraction:

$$A = .001001000011111101101 \ . \ . \ .$$

As before, the probability of stopping with a decision for A is the probability that you get a tail (0) on a flip whose corresponding digit is 1. This probability is equal to the binary fraction itself, because the fraction is expressing the probability as the sum of an endless series of binary fractions, each a reciprocal of a power of 2. And the probability of stopping with a decision for B is the probability you will get a head (1) on a flip whose corresponding digit is 0.

In the first case the probability is $1/2^3 + 1/2^6 + 1/2^{11} + ...$ (The superscripts are the positions of the 1's in the binary fraction for A.) The sum is .00100100001 . . . , the binary fraction for A.

In the second case the probability is $1/2 + 1/2^2 + 1/2^4 + ...$ (The superscripts are the positions of the 0's in the binary fraction for A.) The sum is .11011011110 . . . , which is the complement of the previous fraction; that is, 1's have been replaced by 0's, and 0's by 1's. It is the binary fraction for B.

It is not hard to see why the method works. Probability A is expressed as a sum of an endless series of probabilities. Each is a disjoint event, so their sum

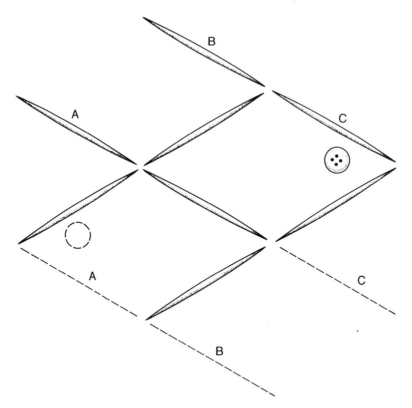

Figure 106   Reversing the fish

must equal probability A. There is, of course, a rapidly decreasing probability that the flipping will not stop, but this probability is vanishingly small. The sequence continues only as long as flip values keep matching their corresponding digits. On the nth toss the probability of such a match is $1/2^n$, which has zero measure in the endless series. In other words, the procedure is practically certain to stop, usually quite soon, with a decision.

7. Once you guess that the stranger is a card—and what a card!—the rest is easy. "Thin, flat voice" is the first hint. The dialogue eliminates first the Joker, then the suits of hearts, diamonds and clubs. A lack of spots makes the stranger a face card, and being unmarried eliminates the king and queen. Only the jack of spades is left.

8. The fish swims the other way if you move three toothpicks and the button as shown in Figure 106.

Sharon Cammel and Jonathan Schonsheck, in a joint letter, pointed out that there are just two ways to move three toothpicks and make the fish swim the other way: one way sending it a trifle higher in the water, the other sending it lower. Tom Kellerman, age eight, found that by moving two toothpicks he could make the fish swim either up or down, although the fish became shorter and fatter.

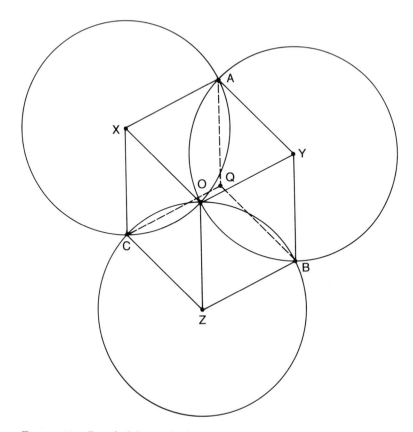

Figure 107 Proof of the circle theorem

9. Draw three line segments connecting O to each of the centers, X, Y and Z, of the three circles. Add six line segments to connect each center to the two nearest of the intersection points, A, B and C. The nine line segments are shown in black in Figure 107. Each line segment is a unit in length; therefore the lines form three rhombuses. Now through each of the intersection points A, B and C draw another line [dotted], making each line parallel to a radius line segment of one of the circles. This forms three additional rhombuses. Because opposite sides of parallelograms are equal, we know that the three dotted line segments are equal and that each is one unit in length. Consequently they meet at a point, Q, that is the center of a circle with a radius of one unit. The intersection points A, B and C lie on this circle, which is the assertion we were asked to prove.

Many readers sent other ways of proving the theorem. For discussions of the problem see George Polya, *Mathematical Discovery,* Vol. 2, 1965, pages 53–58; and Ross Honsberger, Mathematical Gems II, the Mathematical Association of America, 1976, page 18.

CHAPTER SIXTEEN

Look-See Proofs

There is no more effective aid in understanding certain algebraic identities than a good diagram. One should, of course, know how to manipulate algebraic symbols to obtain proofs, but in many cases a dull proof can be supplemented by a geometric analogue so simple and beautiful that the truth of a theorem is almost seen at a single glance.

Consider, for example, a basic summation identity: The sum of the first n positive integers is half of $n(n + 1)$. In equation form,

$$1 + 2 + 3 + 4 + \ldots + n = \frac{n(n + 1)}{2}$$

The first n consecutive positive integers can be depicted by dots in triangular formation [see Figure 108]. Two such triangles fit together to form a rectangular array containing $n(n + 1)$ dots. Because each triangle is half of the rectangle, we see at once that the formula for the number of dots in each triangle is half of $n(n + 1)$.

This simple proof goes back to the ancient Greeks. They called any number of the form $\frac{1}{2}n(n + 1)$ a triangular number, and any number of the form n^2 a square number because it could be represented by a square array of dots. Figure 109, left, shows how square arrays prove that the sum of the first n odd integers is n^2. Think of the pattern as extending any desired distance to the right and down. Each reversed L-shaped strip contains the odd number of circles indicated at the top. It is obvious that each additional strip, that is, each new odd number in the series $1 + 3 + 5 \ldots$, enlarges the square by one unit on a side and that the total number of dots in each square bounded by the nth odd number is n^2.

The Greeks also used square arrays to establish the identity $1 + 2 + 3 + \ldots + n + \ldots + 3 + 2 + 1 = n^2$. The case for $n = 5$ is shown in Figure 109, right. Is any explanation necessary?

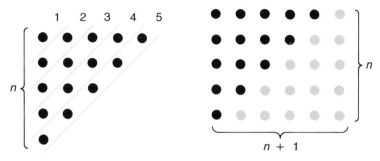

Figure 108 $1 + 2 + 3 + \ldots + n = (n+1)/2$

Finding a formula for the sum of the squares of the first n integers takes a bit more doing. Consider the squares of the first five integers. As we have seen, any square can be represented as the sum of consecutive odd integers starting with 1 *[see Figure 110]*. In these arrays a row of nine dots occurs once, a row of seven dots twice, five-dot rows three times, three-dot rows four times and one-dot rows five times. The 15 rows can be stacked, beginning with the longest on the bottom, to form a skyscraper. By placing square arrays for 1^2, 2^2, 3^2, 4^2 and 5^2 on each side of the skyscraper, we can make a rectangle with a height equal to the sum of the first n integers. As we have seen, this sum is $\frac{1}{2}n(n+1)$. The width of the rectangle is $2n + 1$. The total number of dots in the rectangle is the product of height and width:

$$\frac{n(n+1)(2n+1)}{2}$$

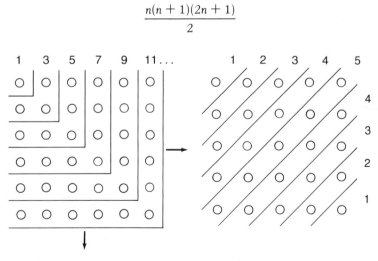

Figure 109   Sum of first n odd integers is n^2
$1 + 2 + 3 + 4 + 5 + 4 + 3 + 2 + 1 = 5^2$

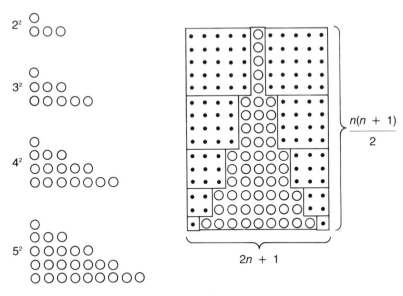

Figure 110 $1^2 + 2^2 + 3^2 + \ldots + n^2(n+1)(2n+1)/6$

The skyscraper, which represents the sum of the squares of the first n numbers, is one-third of the rectangle. Dividing the above formula by 3, therefore, gives the formula for the skyscraper, which is the formula we seek:

$$\frac{n(n+1)(2n+1)}{6}$$

The formula should be familiar to all students of recreational mathematics. It gives the number of different squares, of all sizes, that can be found on a chessboard with n cells on a side. The standard 8-by-8 board, for example, contains $8(8+1)(16+1)/6 = 204$ different squares. It is not hard to see that the formula applies. An 8-by-8 square appears only once on the board. If a 7-by-7 square is placed on the board, it can be shifted to $2^2 = 4$ positions. A 6-by-6 square can be shifted to $3^2 = 9$ positions (eight on the border and one in the center), a 5-by-5 to $4^2 = 16$ positions and so on.

The sum of the cubes of the first n integers is involved in a remarkable identity that astounds most students when they first encounter it. The sum of the first n cubes equals the square of the sum of the first n integers. In algebraic form,

$$1^3 + 2^3 + 3^3 + \ldots + n^3 = (1 + 2 + 3 + \ldots + n)^2.$$

An old diagram for it is shown in Figure 111. The square array of numbers,

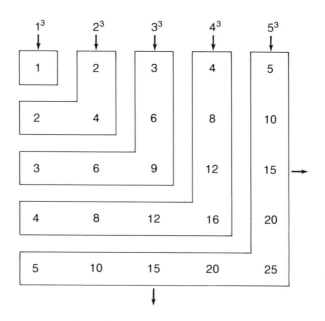

Figure 111 $1^3 + 2^3 + 3^3 + 4^3 + 5^3 = (1 + 2 + 3 + 4 + 5)^2$

which extends down and right to infinity, is simply the multiplication table. Each number is the product of the number at the left of its row and the number at the top of its column. The table is divided into bent strips, and the sum of the numbers in each nth strip is n^3. With a square of five bent strips the sum of all the numbers is $1^3 + 2^3 + 3^3 + 4^3 + 5^3$. Since this square is the multiplication table through 5, it is equally clear that the sum of all the numbers is $(1 + 2 + 3 + 4 + 5)(1 + 2 + 3 + 4 + 5)$, or $(1 + 2 + 3 + 4 + 5)^2$.

Unfortunately this geometric analogue is not as good a "look-see" proof as the preceding examples are. It is not instantly obvious that the numbers in each nth bent strip have a sum of n^3. A more elegant geometric analogue of the same identity was devised by Solomon W. Golomb and published in 1965. The isomorphism [*see Figure 112*] is easily explained. The large square has a side that equals the sum of the first eight integers, so its area is

$$(1 + 2 + 3 + 4 + 5 + 6 + 7 + 8)^2.$$

This gives one side of the identity. For the other side note that the large square is made up of one square of side 1, two squares of side 2, three of side 3, four of side 4 and so on up to eight squares of side 8. For squares of even sides there is a square overlap, shown in black, but each overlap is adjacent to an empty square region, shown white, which is the same size. We can therefore take one of each pair of black overlapping squares and use it for plugging the hole next to it and,

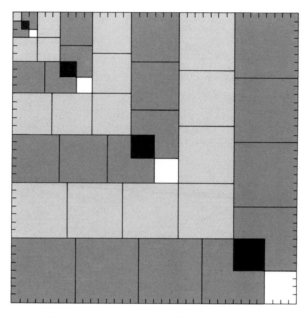

Figure 112 $1^3 + 2^3 + 3^3 + \ldots + n^3 = (1 + 2 + 3 + \ldots + n)^2$

in this way, eliminate all overlaps and holes. Now, $1 \times 1^2 = 1^3$, $2 \times 2^2 = 2^3$, $3 \times 3^2 = 3^3$ and so on. In other words, the total area is $1^3 + 2^3 + 3^3 + 4^3 + 5^3 + 6^3 + 7^3 + 8^3$, which is the other side of the identity.

In the same article Golomb provides another proof, based on a suggestion by Warren Lushbaugh, for the same summation identity *[see Figure 113]*. The squares shown have sides 1, 2, 3, 4 and 5. There are no holes or overlaps. Each square of side n appears $4n$ times. We can write the identity

$$4(1 \times 1^2 + 2 \times 2^2 + 3 \times 3^2 + \ldots + n \times n^2) = [2(1 + 2 + 3 + \ldots + n)]^2$$

which simplifies to the same identity as before. The sum of the first n integers is $\frac{1}{2}n(n + 1)$, and since the square of this equals the sum of the first n cubes, we can represent the sum of the first n cubes by the compact formula

$$\left[\frac{n(n + 1)}{2} \right]^2$$

This too is a formula that puzzlists should know. It not only gives the number of different cubes, of all sizes, contained in a cubical chessboard of side n but also gives the number of different rectangles of all sizes (including squares) in a flat chessboard of side n. Thus the standard chessboard of side 8 contains 1,296

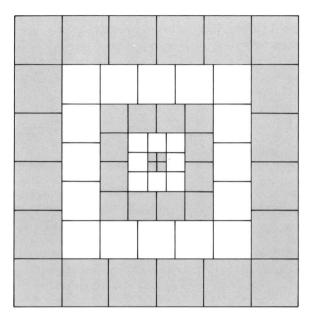

Figure 113 $1^3 + 2^3 + 3^3 + \ldots + n^3 = (1 + 2 + 3 + \ldots + n)^2$

rectangles, and a three-dimensional board of side 8 contains 1,296 cubes. We can "see" how this counts the cubes by the same mechanical argument we applied to the squares of a chessboard. There is one largest cube of side 8. An order-7 cube goes in $2^3 = 8$ corners, an order-6 cube in $3^3 = 27$ spots and so on.

The flat board provides still another geometric way of proving that the sum of the first n cubes equals the square of the sum of the first n integers. Robert G. Stein explained in 1971 how two counting arguments for the number of rectangles in a square chessboard give the two sides of the identity. See also Gene Murrow's more detailed solution of the rectangle-counting problem in his 1971 article.

We turn now to another class of geometric analogues: dissections that illustrate simple identities involving squares that are the sums of other squares and cubes that are the sum of other cubes. Take, for instance, the familiar Pythagorean triplet $3^2 + 4^2 = 5^2$. It is the only such triplet of consecutive positive integers. How can we cut an order-5 square into the fewest number of polyominoes that can be rearranged to make two squares of sides 3 and 4? Two solutions in four pieces — one with the 3-square intact, the other with the 4-square intact — are shown in Figure 114, top. The dissection cannot be done with fewer pieces. No polyomino can be longer than four units; therefore the 5-square must be divided by a cut joining left to right sides and by another cut

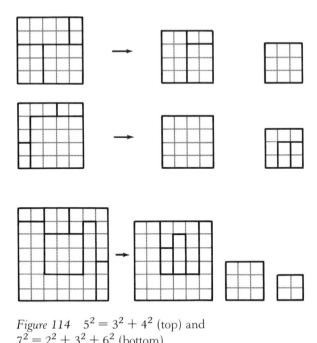

Figure 114 $5^2 = 3^2 + 4^2$ (top) and
$7^2 = 2^2 + 3^2 + 6^2$ (bottom)

joining top to bottom — a procedure that must produce at least four pieces. Because there are many four-piece solutions, recreational geometers amuse themselves by adding other provisos. In the two solutions shown, the total cutting length (10 units) is minimal, and in the top solution all the polyominoes are rectangles.

Henry Ernest Dudeney's puzzle books contain many dissection problems that illustrate other square identities. For example, the solution of Problem 357 in *536 Puzzles and Curious Problems* is an analogue of $2^2 + 3^2 + 6^2 = 7^2$. The pattern has six pieces and a cutting length of 27 *[see Figure 102, bottom]*. Can the reader find a better solution with only five polyominoes?

Two cubes cannot have a cubical sum, but $w^3 + x^3 + y^3 = z^3$ has an infinity of integral solutions. The only solution in consecutive positive integers (indeed, the only solution with the first three terms consecutive) is $3^3 + 4^3 + 5^3 = 6^3$. This suggested to the British mathematician John Leech the following pretty problem: How can the order-6 cube be cut along integral lattice planes into a minimum number of polycubes (pieces formed by joining unit cubes) that will make separate cubes of sides 3, 4 and 5?

E. H. Wheeler was the first to solve it. His eight-piece solution was published in *Eureka* (an annual publication of the Archimedean Mathematical Society of the University of Cambridge), Volume 14, 1951, page 23. In Wheeler's dissec-

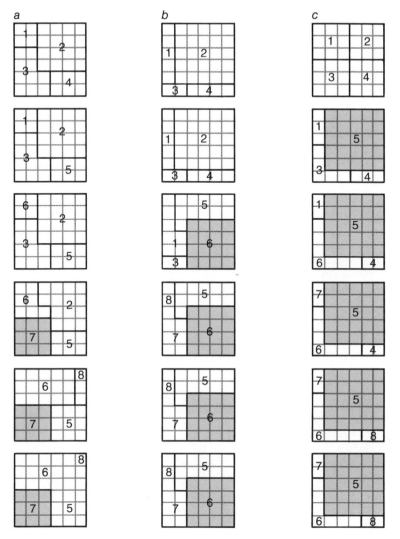

Figure 115   Order −6 cube's dissection by Wheeler (a), O'Beirne (b) and Duffy (c)

tion, shown by the six cross sections in the left column of Figure 115, the 3-cube remains intact. (Cross sections of the intact cubes are shaded in all three solutions.) A simpler eight-piece solution was later found by J. H. Thewlis of Argyll, Scotland, which Thomas H. O'Beirne of Glasgow further simplified as shown in the illustration's middle column. The 4-cube remains intact, and only two polycubes are not rectangular blocks. A remarkable eight-piece dissection, with the 5-cube intact, found in 1970 by Emmet J. Duffy of Oak Park, Ill., is shown in the column at the right.

O'Beirne asked himself: What is the minimum number of pieces for a solution of this problem in which all the polycubes are "blocks" (rectangular

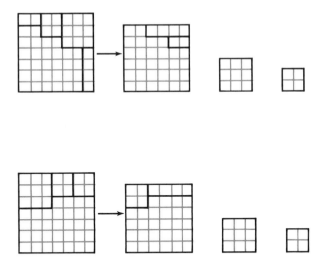

Figure 116   Five-piece dissections for $7^2 = 6^2 + 3^2 + 2^2$

parallelepipeds)? It proved to be a difficult question. There must, of course, be at least eight polycubes, regardless of their shapes. No polycube can be six units long in any direction. This requires that the 6-cube be divided by at least three intersecting slices: one cutting all left-to-right rows, one cutting all front-to-back rows and one cutting all top-to-bottom columns. The procedure produces at least eight polycubes. By more complicated reasoning O'Beirne was able to show that an eight-piece dissection is not possible with all rectangular blocks.

Is it possible with nine blocks? Yes, In 1971 O'Beirne obtained a nine-block dissection that he believes is unique, although he has not been able to prove it. It is difficult to find the nine-block dissection without using a computer program. O'Beirne's solution is given in the answer section.

In working on this polycube problem and others, it is helpful to have a few hundred plastic interlocking cubes of different colors. Such cubes are available from firms that sell mathematical supplies to teachers.

ANSWERS

The first problem was to improve on Dudeney's dissection of a 7-square into six pieces that form squares of sides 2, 3 and 6. At the top of Figure 116 is shown a five-piece dissection. I thought this was unique in having the minimal cutting length of 16 units, but Graham Lord sent the five-piece dissection shown at the bottom of Figure 116 that also has a cut length of 16.

Figure 117 shows in six cross sections how Thomas H. O'Beirne sliced a 6-cube into nine rectangular blocks (the minimum) that can be reassembled to make separate cubes of sides 3, 4 and 5. No two "bricks" in this remarkable

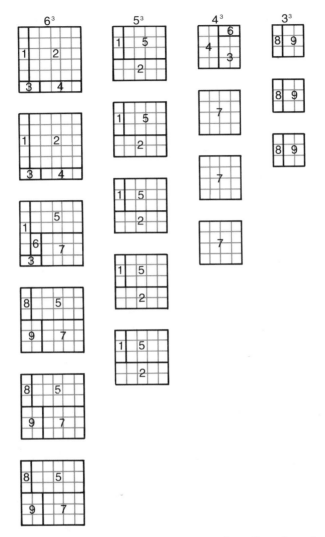

Figure 117 Nine-block dissections for $6^3 = 5^3 + 4^3 + 3^3$

dissection are alike. O'Beirne has shown that a dissection in eight blocks is not possible, but it is not known whether his nine-piece dissection is unique.

ADDENDUM

Alistair J. McIntosh wrote from England to comment on the multiplication square shown in Figure 111. If we draw on its matrix any rectangle that has its top left cell in the top left corner of the matrix, the number in the bottom right corner of the rectangle gives the number of cells in the rectangle. Knowing this, further inspection of the table at once discloses a variety of arithmetical truths not otherwise obvious. For example, we see that multiplication is commutative because an *m*-by-*n* rectangle has the same number of cells as an *n*-by-*m* rectan-

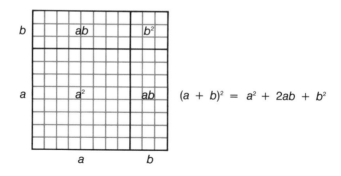

$(a + b)^2 = a^2 + 2ab + b^2$

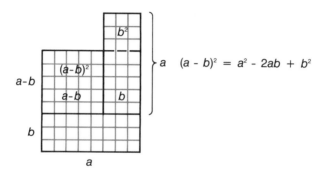

$(a - b)^2 = a^2 - 2ab + b^2$

Figure 118   Visual proofs of two identities

gle. We see why square numbers are called square and why the square numbers must be on a main diagonal of the table. If we move one step diagonally up and right from any square number, we hit a number one less than the square. This clearly is the same as comparing the area of a square with the area of a rectangle having one side that is one less than the side of the square and another side that is one more than the side of the square. In algebraic terms we have uncovered the identity $n^2 - 1 = (n + 1)(n - 1)$. Many other algebraic equations can be understood by carefully studying the matrix.

I did not have space in my column for mentioning how easy it is to diagram the familiar identities $(a + b)^2 = a^2 + 2ab + b^2$ and $(a - b)^2 = a^2 - 2ab + b^2$.

The diagrams in Figure 118 are look-see proofs of these two equations. Although these diagrams are as old as algebra, it is surprising how few teachers bother to display them to students. Figure 119 shows how readily this type of diagram can be extended to three dimensions to display the cubic equation $(a + b)^3 = a^3 + 3a^2b + 3ab^2 + b^3$. If I were teaching algebra, I would have this model available for students to take apart to verify the identity by calculat-

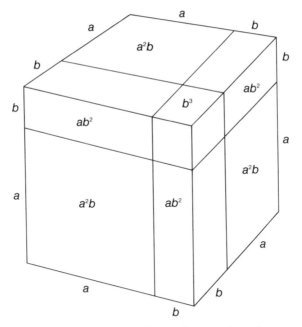

Figure 119 $(a + b)^3 = a^3 + 3a^2b + 3ab^2 + b^3$

ing the volumes of the eight polycubes and adding them to get the volume of the cube they form. (In the picture, only the a^3 piece is not visible.)

If we do not require that the two smaller squares, in a dissection of the 5-square to a 3-square and a 4-square, be separated but allow them to be joined, the $3^2 + 4^2 = 5^2$ equality can be displayed by a dissection of just three pieces, as shown in Figure 120. Similar reductions in the number of pieces can be made for other models of square and cubic identities. Consider the slicing of the 6-cube into polycubes that will form cubes of sides 3, 4 and 5. If the three smaller cubes are allowed to be attached, E. J. Duffy found scores of solutions in as few as six pieces, including many that produce a neat tower of the 3-cube on the 4-cube on the 5-cube. Can a solid joining the three cubes be produced by cutting the 6-cube into fewer than six polycubes? This question remains unanswered.

A famous anecdote tells how G. H. Hardy, visiting the East Indian mathematician Ramanujan in a hospital, remarked that the number of his cab, 1729, was a dull number. On the contrary, Ramanujan promptly replied, it is the

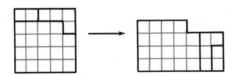

Figure 120   A 3-piece proof of $3^2 + 4^2 = 5^2$

smallest integer that can be expressed as the sum of two cubes in two different ways: $1729 = 1^3 + 12^3 = 9^3 + 10^3$. In 1970 J. H. Cadwell published a paper (see the bibliography) showing how a $7 \times 13 \times 19 = 1,729$ block can be cut into 12 polycubes that will form a pair of cubes with sides 1 and 12 and a pair of cubes with sides 9 and 10.

BIBLIOGRAPHY

"Geometrical Arithmetic." Helen A. Merrill in *Mathematical Excursions,* Chapter 10. Dover, 1957.

Mathematical Models, 2nd edition. H. Martyn Cundy and A. P. Rollett. Oxford University Press, 1961.

Vision in Elementary Mathematics. W. W. Sawyer. Penguin, 1964.

"Some Dissection Problems Involving Sums of Cubes." J. H. Cadwell in *Mathematical Gazette,* Vol. 48, 1964, pages 391–396.

"A Geometric Proof of a Famous Identity." S. W. Golomb in *Mathematical Gazette,* Vol. 69, 1965, pages 198–200.

536 Puzzles and Curious Problems. H. E. Dudeney. Scribner's, 1967.

"A Combinatorial Proof that $\Sigma k^3 = (\Sigma k)^2$." Robert G. Stein in *Mathematics Magzazine,* Vol. 44, 1971, pages 161–162.

"A Three-way Dissection Based on Ramanujan's Number." J. H. Cadwell in *Mathematical Gazette,* Vol. 54, 1970, pages 385–387.

"A Geometric Application of the 'Shepherd Principle.'" Gene Murrow in *Mathematics Teacher,* Vol. 64, 1971, pages 756–758. ("Shepherd principle" is defined as follows: to count sheep, count the number of their legs and divide by four.)

"Geometric Solutions to Quadratic and Cubic Equations." Harley B. Henning in *Mathematics Teacher,* Vol. 65, 1972, pages 113–119.

"A Physical Model for Factoring Quadratic Polynomials." James K. Bidwell in *Mathematics Teacher,* Vol. 65, 1972, pages 201–205.

"Another Geometric Introduction to Mathematical Generalization." H. L. Kung in *Mathematics Teacher,* Vol. 65, 1972, pages 375–376.

CHAPTER SEVENTEEN

Worm Paths

Fashionable methods of teaching mathematics to children come and go. (The last to go was the "modern math" fiasco.) One of these days mathematics teachers will discover what John Dewey tried to tell them 75 years ago: Children learn best by doing something they enjoy. With this in mind, Seymour A. Papert, a former assistant to Jean Piaget, who is now working in the Artificial Intelligence Laboratory of the Massachusetts Institute of Technology, has designed a variety of animal robots that can be controlled by a desk computer. One of them is a "turtle" with a pen on its underside. Suitably programmed, the turtle draws geometric figures by crawling across large sheets of paper on the floor.

Defining geometric figures as paths generated by a moving point is an ancient idea. Consider a square. Instead of calling it a four-sided polygon with equal sides and angles, call it the path traced by a worm crawling over a plane according to the following rule: Go straight for distance k, turn 90 degrees left and repeat until the path returns to its origin.

The idealized worm (moving point) can obviously be programmed to generate any pattern of lines. A challenging recreational task now presents itself. What kinds of program, with extremely simple rules, give interesting or beautiful patterns? A good way to simplify rules is to restrict the worm to paths along a regular lattice. This enables one to experiment with rules by drawing paths on square or isometric graph paper. Better still, if one has access to a computer with a display screen, one can write simple programs and then enjoy the spectacle of watching a path of light grow on the screen.

Frank C. Odds, a British biochemist, recently proposed a class of rules for generating patterns that he named spirolaterals. The worm crawls a distance of one unit, turns, crawls two units, turns, crawls three units and so on, traversing distances in counting order until the length of a path segment reaches a specified integer, n, when the procedure starts over. The turning angle is always the

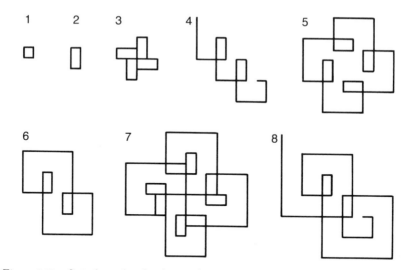

Figure 121   Spirolaterals of orders 1 through 8 with all turns in same direction

same, but the directions of turn may be right or left according to the worm's program. The number n, which is the number of segments in the counting series and also the number of turns before the series repeats, is called the order of the spirolateral.

Two examples will make this clear. If the order is 1, the angle is 90 degrees and all turns are in the same direction, the spirolateral is a square. If the order is 7, the angle is 90 and all turns are the same, the spirolateral has a pleasing closed pattern. Figure 121 shows all the right-angle spirolaterals of orders 1 through 8 when all turns are the same. It is easy to see why Odds chose the name spirolateral: "lateral" for flat surface, and "spiro" for the square spirals that generate the figures.

Note that spirolaterals of orders 4 and 8 do not close (return to the origin). As Odds puts it, they meander jerkily to infinity. Indeed, it is not hard to prove that there is no closure for orders in the series 4, 8, 12, 16, . . . , that two repetitions of the square spiral will close the figure (producing twofold symmetry) if the order is in the series 2, 6, 10, 14, . . . and that all other orders close after four repetitions and have fourfold symmetry.

There is no reason why all turns must be the same way, but when they vary the situation becomes complicated enough to call for a compact notation. Odds suggests writing the angle as a subscript of the order number and using a superscript on the left or the right to show which turns are left and which right. For example, $^{1,7}9_{90}$ defines the spirolateral generated by nine turns of 90 degrees, with left turns preceding segments 1 and 7 and (by implication) all the

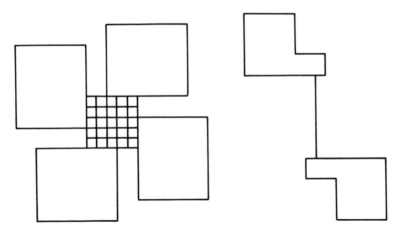

Figure 122 Ninety-degree spirolaterals with mixed turns: 9_{90}^{6} (left) $8_{90}^{1,4,8}$ (right)

other segments turning right. The same spirolateral could be defined by putting a superscript on the right to indicate right turns, $9_{90}^{2,3,4,5,6,8,9}$, but the first notation seems preferable.

Figure 122 shows two right-angle spirolaterals with mixed left and right turns. Figure 123 shows spirolaterals with angles of 36, 45 and 60 degrees. Spirolaterals with 60-degree turns can be easily drawn on isometric paper. Those with 45-degree angles are not hard to construct with a ruler and a draftsman's triangle, but those with other angles take more doing. Any angle that is an exact divisor of 180 degrees will generate a spirolateral.

Not much is known about spirolaterals. Sometimes two different rules pro-

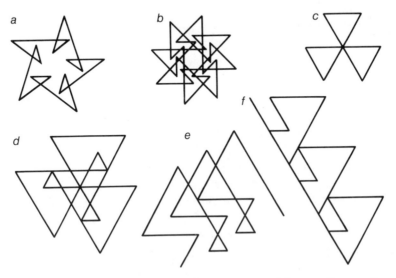

Figure 123 Spirolaterals (a) 3_{36} (b) $^{1}3_{45}$ (c) 2_{60} (d) $^{1,2}5_{60}$ (e) $^{3}5_{60}$ (f) $^{2}5_{60}$

duce spirolaterals that are mirror images (for example $^{1,2,3}7_{90}$ and $^{5,6,7}7_{90}$), but no one knows how to predict this without drawing the figures. Nor is it known how to tell from looking at a spirolateral's formula, except in special cases, whether or not the figure will close and, if it does, how many repetitions are needed to close it.

A few years ago John Horton Conway of the University of Cambridge suggested a new approach to worm paths. Instead of viewing the worm as an explorer, view it as an eater. Food is confined to the lines of an arbitrarily large grid. The worm hatches from an egg at one node and then starts to crawl along grid lines, eating as it goes. A fixed set of rules determines the worm's decision at each node. It is assumed that the worm never traverses a segment already eaten. No segment lengths need to be specified, only the direction of turn, the direction depending solely on the state (eaten or uneaten) of all segments meeting at that node.

"Paterson's Worm," by Michael Beeler, a memorandum issued in 1973 by M.I.T.'s Artificial Intelligence Laboratory, deals entirely with such worm tracks. What follows, either directly quoted or paraphrased, is taken from the memorandum with the permission of Beeler and Marvin L. Minsky, who heads the laboratory.

"Certain prehistoric worms fed on sediment in the mud at the bottom of ponds," Beeler's memorandum begins. "For efficiency, they would not retrace paths which had already been traveled, since little food was left there. Yet food probably occurred in patches, so it was desirable to stay near previous trails. Worms had innate 'rules' regarding how close to 'eaten paths' to stay, how far to go before turning around, how sharp a turn to make, etc. These rules varied from species to species, and paleontologists can trace the development of species and determine the similarity of different species by comparing fossil records of worm tracks. (See *Science* magazine, 21 November 1969, for further details and a discussion of computer simulation of natural worm tracks.)

"Early in 1971, Michael Paterson [a computer scientist at the University of Warwick] mentioned to me a mathematical idealization of the prehistoric worm. He and John Conway had been interested in a worm constrained to eat food only along the grid lines of graph paper. . . .

"If a worm, arriving at a node with no segments eaten (except of course the one it just ate), should find in its rules, 'For this distribution, go straight,' then the worm will go straight forever. Since this is neither interesting to us, nor very useful to a real worm, which would quickly reach the edge of its food patch, we discard it. We require that, upon discovering a virgin node, all sets of rules must say to turn. To avoid mirror-image duplication, we require that the turn be to the worm's right (clockwise as seen from above)."

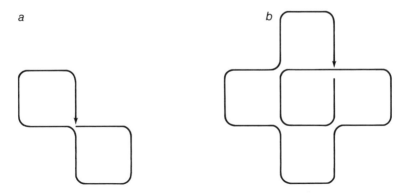

Figure 124   Fossil tracks of the only two species of simple quadrille worms

Consider what Beeler calls a "simple quadrille worm": a worm crawling on a square grid and turning right at each node. What happens after it traces a square? It cannot turn right again because this would take it along an eaten segment. It has only two choices. If it is programmed to turn left if, and only if, it cannot turn right, it will trace two squares *[see Figure 124a]*. No uneaten segments are available, so the worm dies. If it is programmed to go straight when it cannot go right, and left when it cannot go straight or right, it will trace five squares before it expires at its origin *[b]*. These two fossil tracks exhaust the variety of species of simple quadrille worms.

To avoid a worm's early demise, Conway proposed that quadrille worms have the ability to "look ahead" and see the distribution of eaten and uneaten segments at each adjacent node. For example, a worm could be programmed to turn right if, and only if, it senses that this will take it to a node where four uneaten segments meet. Otherwise it goes straight. The result is a simple square spiral. If the rule is to turn left whenever a right turn leads to a node with an eaten segment, the path is a more interesting spiral *[see Figure 125]*. It is easy, of course, to produce more elaborate paths by complicating the rules.

The rules may allow anything. What happens when rules allow look-ahead worms to hop? What happens when barriers are suitably placed or when the grid is bounded on all sides? What happens when two or more worms of the same or different species interact? What happens when a newly hatched worm crawls a short distance along a defined path (such as a straight line of three units) before its repetitive behavior starts? How about two armies of worms crawling toward each other, each army obeying a different program? Are there possibilities here for competitive games? Are there interesting paths or patterns in three dimensions and higher dimensions?

Beeler avoids such difficult questions by confining his attention to what I shall call "simple isometric worms": worms that feed along an isometric grid of

Figure 125 Infinite path of a look-ahead quadrille worm that turns left when it cannot turn right

unit equilateral triangles. They are simple worms because they do not look ahead. With the isometric grid, however, six segments (not four) meet at each node. It seems to be a trivial difference, but, as Beeler makes clear, the possibilities for variant rules allow the definition of no fewer than 1,296 species.

All simple worms, quadrille or isometric, obey three general rules:

1. If no segments have been eaten at a node (except the segment just traveled), the worm turns right.

2. If all segments at the node have been eaten, the worm dies.

3. If only one segment at the node is uneaten, the worm takes it.

As we have seen, a simple quadrille worm following the above rules encounters only one "field" in which it must choose. Since it has only two choices, we can define only two species. The isometric grid, in contrast, offers a simple worm four major fields (one consisting of four subfields) in which decisions must be made. It is the behavior of these simple isometric worms that Paterson became interested in, and Beeler's memorandum also is primarily concerned with their behavior.

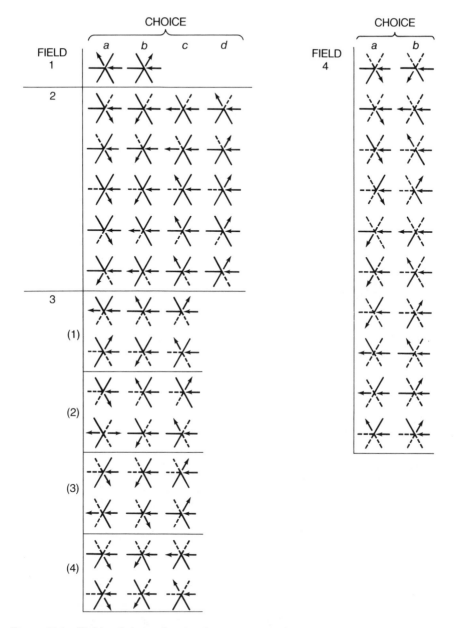

Figure 126   Fields of choice for simple isometric worms

The four major fields and all their choices are shown in Figure 126. The black lines indicate uneaten segments,, the dotted lines are eaten segments of the worm's path and the arrows show how the worm approaches and leaves a node.

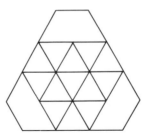

Figure 127   Path traced by 14 simple isometric worms

The four fields are the following:

1.   The worm approaches a node with no eaten segments other than the one it has just traversed. Its right turn can be either "gentle" (120 degrees) or "sharp" (60 degrees). Number of choices: two.

2.   The worm meets one eaten segment as it returns for the first time to its origin. As the chart shows, it can approach this node along any of five different segments. For each approach there is a choice of leaving by one of four uneaten segments. Number of choices: four.

3.   The worm meets two eaten segments as it returns to a point along its path. The node will be the vertex of either a sharp turn or a gentle one. In either case the approach can be made in four ways. For each way the worm can leave by one of three uneaten segments. Number of choices: $3 \times 3 \times 3 \times 3 = 81$.

4.   The worm meets three eaten segments as it returns for the second time to its origin. This can happen in 10 ways, but each requires only a choice between two ways of leaving the node. Number of choices: two.

Meeting four or five eaten segments does not offer the worm a choice. In the first case it must take the only remaining uneaten segment. In the second case it has no segment to take, so it dies. Thus there are $2 \times 4 \times 81 \times 2 = 1,296$ sets of rules, each defining a distinct species of simple isometric worm.

Beeler uses the pattern shown in Figure 127 to explain how the rules work. The notation adopted here (Beeler uses a more compact notation based on octal and binary digits) identifies this as a fossil path generated by a $1_a 2_b 3_{acac} 4_b$ worm. The formula tells us that when the worm faces a choice in Field 1, it follows Rule a. (It makes a gentle, not a sharp, turn.) For Field 2 it chooses b. The four subscripts after 3 refer to the four subfields of Field 3, where choices a, c, a, c are in the worm's feeding program. Finally, for Field 4 the worm selects b. As Beeler suggests, the reader might pause at this point to see if he can trace the path on a sheet of isometric paper, following the seven rules given by the

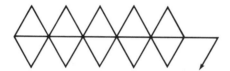

Figure 128 Zipper path of 54 sharp worms

formula. (It is necessary at each node to rotate the paper until the pattern at that node corresponds to the proper diagram on the chart.)

The path just described is generated by 14 worms. Some trace it exactly the same way. For example, the choice presented by the first subfield of Field 3 never arises; therefore it does not matter which of the three choices, *a, b* or *c,* is in the program. We indicate this in the formula by putting the three choices inside parentheses: $1_a2_b3_{(abc)}cac4_b$. The formula now describes three different worms. Each draws the same path the same way. Other worms draw the same path in different ways.

Beeler's computer program explored the behavior of all 1,296 species of simple isometric worms. Results show that 209 species generate paths that are unique, in the sense that no other species generates them. Forty-six paths are each characteristic of two species, and 44 paths belong to more than two. Thus there are 299 distinct paths all together.

The simplest of the 299 paths is the radioactivity symbol *[see Figure 123c].* It has the smallest length (nine units) and also the fewest number of nodes (seven). It is the path traced by the largest number of worms, no less than 162 species. Using parentheses to show alternatives, we obtain one formula, which defines all 162:

$$1_b2_{(ac)}3_{(abc)(abc)(abc)(abc)}4$$

The formula shows at once that it describes $2 \times 3 \times 3 \times 3 \times 3 = 162$ sets of rules. If the reader will test this formula on isometric paper, he will find that, regardless of the choice he makes inside each pair of parentheses, the radioactivity symbol will result.

Is there a longest path? No, because many paths never return to their origin a third time and are therefore infinite. A trivial infinite path of a type called the zipper is produced by 54 sharp worms with the formula $1_b2_d3_{(abc)(abc)}b4_{(ab)}$ *[see Figure 128].*

Other worms spiral forever around their origin. The spiral shapes, can be hexagons, diamonds, triangles, stars and asymmetric shapes such as the one shown in Figure 129. "Shoot growers," another class of infinite paths, start in a conventional way, and then the worm falls into a curious spiraling action that

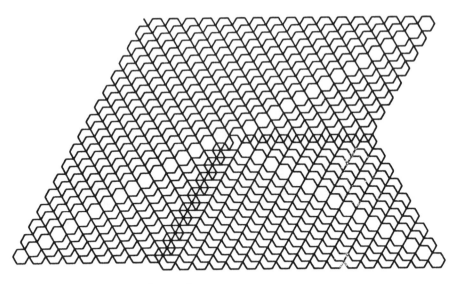

Figure 129   Three-pointed spiral by a $1_a2_c3_{acba}4_a$ worm

keeps repeating itself with regular displacements, creating a rod that shoots off to infinity *[see Figure 130]*.

The longest of the known finite paths is shown in Figure 131. The path is generated by a $1_a2_d3_{cbac}4_b$ worm and has 220,142 units. Note the crystalline regularity of the border and the lines that crisscross near the border, and how strikingly this symmetry contrasts with the dishevelment near the center. The point of origin, not identifiable, is at a spot just to the left of the center of the pattern.

One speaks of "known" finite paths because about a dozen worms have paths so long that no one yet knows whether they are finite or infinite. The path generated by $1_b2_a3_{bcaa}4_b$, for instance, was tracked by Beeler to a length of 10 million units without revealing whether the worm dies or goes on gorging itself forever.

Some path patterns have outlines as unstructured as a cloud *[see Figure 132]*. Others, such as the "superdoily," display the rigid sixfold symmetry of a snow crystal *[see Figure 133.]* Note the six-pointed star in the center. Occasionally one such star or more appear as randomly situated white spots inside a gray plenum of closely packed unit triangles. Sometimes the stars are separated; sometimes they are partially merged to form binary or ternary systems. The situation is not unlike that of the physical world, in which low-order mathematical laws generate a rich assortment of both patterned and unpatterned objects, as well as objects in which order and disorder, symmetry and asymmetry are crazily mixed.

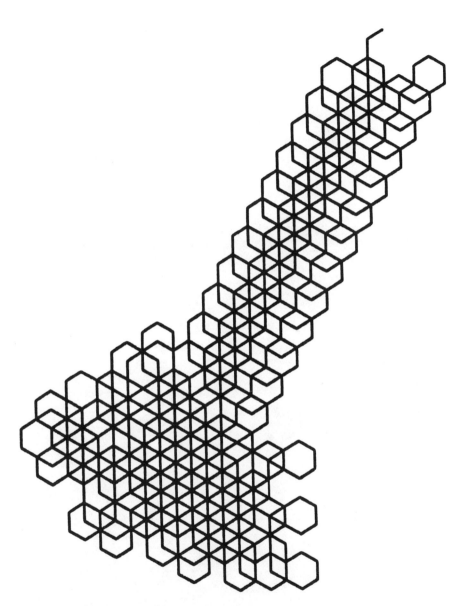

Figure 130   Shoot grower by a $1_a2_d3_{caaa}4_b$ worm

Beeler's work clearly has just scratched the surface of isometric-worm pathology. His program investigated only the simplest genus, with no attempt to complicate the genetic rules or to find out what happens when simple worms interact with barriers, boundaries or other worms of the same or different species.

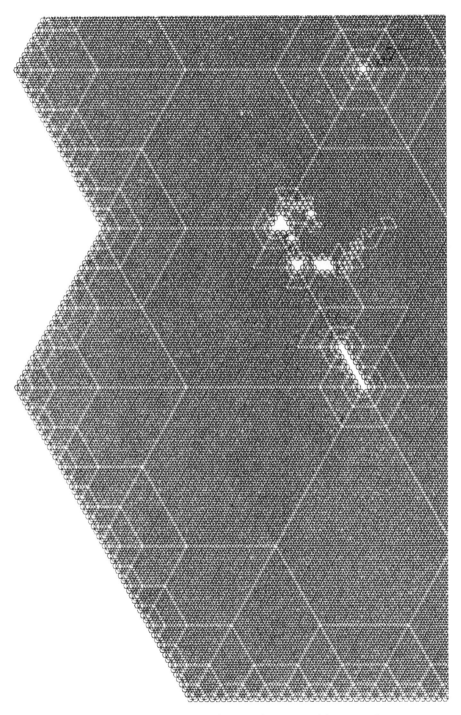

Figure 131 Part of the longest known finite path generated by a simple isometric worm. The entire path is about twice as large as the segment shown on this page.

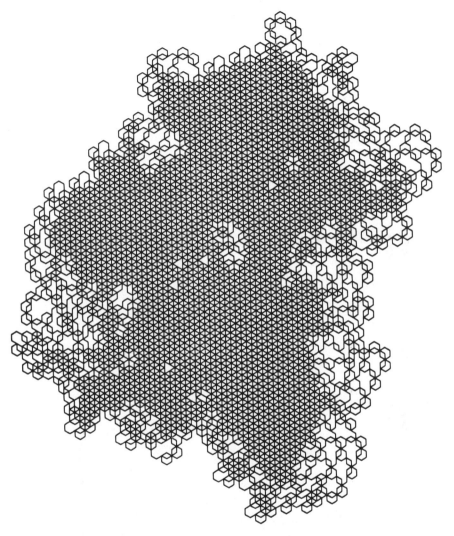

Figure 132   Cloud path generated by a gentle worm $1_a2_{abaa}3_c4$

On August 13, 1973, *Newsweek* ran a story about Papert's turtle. A girl who had ranked near the bottom of her mathematics class was programming the turtle to draw a certain design. "That math must be fun," said a passing visitor. "There ain't *nothin'* fun in math," the girl replied. She had no idea she was doing math, reported *Newsweek*, and Papert saw no reason to tell her.

Some of the questions raised about "spirolaterals" have been answered by readers. The main problem — How can you determine from a spirolateral's

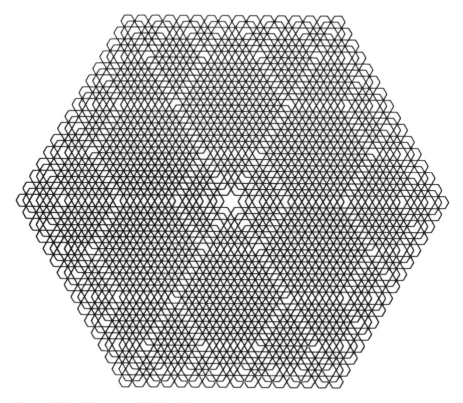

Figure 133   The superdoily $1_a2_d3_{cbaa}4_b$

formula whether it closes and, if it does, in how many repetitions? — was solved by James Thomas, William Laubenheimer, Steven Wolfson and E. Lawrence McMahon. It turns out (as other readers also reported) that a spirolateral results whenever the angle of turn is a rational number. If the angle is irrational, the spirolateral remains within a bounded region but never closes.

Thomas gave the following procedure: First determine the angle's supplement (its difference from 180 degrees). Multiply this by the difference between the number of left turns and number of right turns. (The difference is equal to the spirolateral's order minus twice the number of left turns.)

From the above result, keep subtracting 360 until the remainder is between -180 and 180. Take the absolute value and call it x. This represents the net angular change after each cycle. If x equals 0, there is either no closure (and the spirolateral is infinite) or it closes after the first cycle.

If x is not zero, divide it into the lowest multiple of 360 that it will go into evenly. The result is the number of cycles required to close the spirolateral.

To express this procedure by a compact formula, McMahon proposed letting n equal the spirolateral's order, k equal the number of left turns and m equal 360 divided by the rational angle. Write the fraction

$$\frac{(m-2)(n-2k)}{2m}$$

and reduce to lowest terms. If the result is an integer, the spirolateral either does not close or closes after one cycle. If the result is an integral fraction, a/b, the figure closes after b cycles.

Seymour Papert of MIT now heads his own firm, Logo Systems, Inc., in Montreal, where he continues to explore and develop ways of teaching math to children by way of computer systems that use an extremely flexible language called LOGO, which he invented. The original mechanical turtle that crawled over the floor has become a "turtle" symbol on the computer screen. Manipulated by a keyboard or a joystick or a mouse, the little beast traces patterns on the screen. Not only does the turtle teach geometry, but by giving it velocity and acceleration — Papert calls the enhanced reptile a "dyna-turtle" — it can teach elementary physics. You'll find all this explained in Papert's delightful book *Mindstorms: Children, Computers, and Powerful Ideas*, published by Basic Books in 1980.

David Maynard, a computer hacker with Electronic Arts, in San Mateo, Calif., read my column on worms and was inspired to invent an exciting computer game in which two, three or four worms battle one another on the screen. The game was introduced by Electronic Arts in 1983 as WORMS, one of several games on a floppy disk called "Golden Oldies" that runs on a Commodore or Atari. The triangular grid is toroidal — top and bottom, and left and right sides, wrap around. Each worm is a finite-state machine programmed by its player. When a worm moves, a sound output identifies the worm and which way it is moving. When a node is eaten, it changes to a color that indicates which worm ate it. The player whose worm eats the most nodes is the winner. The game is too complex to analyze but great fun to play, and players soon develop intuitive skills. In its December 1983 issue, *Omni* selected WORMS as one of the year's 10 best games.

Michael Beeler, who now works for a computer firm in Cambridge, Mass., tells me that, so far as he knows, all the worms listed in his memo as "uncertain" as to whether they are mortal or immortal are still uncertain. The memo is out of print, but copies can be purchased from NTIS Information Center, Springfield, Va., or from M.I.T. Microproduction Laboratory.

A letter from Geoffrey Wyvill, in England, described an amusing worm path. Consider the counting numbers as they are represented in binary notation. Assign to each number a 1 or a 0, depending on whether there is an odd or

Figure 134 Samples of R. L. Phillips's complex spirolaterals

an even number of 1's in the number. This generates the sequence 1,1,0,1,0,0,1,1,0,0,1,0, ... Give this sequence to a worm, telling it to go straight ahead one step at each 0 and to turn left at each 1 and go one step. Wyvill expected the path to be "large and rambling." To his surprise, the worm never left the closed path shown below:

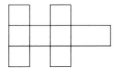

I passed Wyvill's letter along to Beeler, who had no difficulty proving that the poor worm is permanently trapped.

R. L. Phillips, an engineer at the University of Michigan, wrote a computer program for generating complex spirolaterals. He sent me a large number of sample printouts, from which I have selected the six shown in Figure 134.

BIBLIOGRAPHY

"Spirolaterals." Frank C. Odds in *Mathematics Teacher,* Vol. 66, 1973, pages 121 – 124.

Paterson's Worm: Artificial Intelligence Memo No. 290. Michael Beeler. Massachusetts Institute of Technology, June, 1973.

"Spirolaterals: An Advanced Investigation from an Elementary Standpoint." Alice Kaseberg Schwandt in *Mathematics Teacher,* Vol. 72, 1979, pages 166 – 169.

"Square Spirolaterals." Margaret Kenney and Stanley Bezuszka in *Mathematics Teaching,* Vol. 95, 1981, pages 22 – 27.

"Spirolateral." No byline. *Student Math Notes,* September, 1983.

CHAPTER EIGHTEEN

Waring's Problems

In 1909 David Hilbert, the noted German mathematician, proved a lucky guess about numbers that had been made 139 years earlier by the English algebraist Edward Waring. Hilbert's proof, one of his great achievements, has led to significant developments in number theory and has raised a myriad fascinating related questions. Many of the questions, unsolved to this day, are so easy to understand, and so amenable to amateur empirical research by anyone with tables of powers and a desk computer, that perhaps a reader of this chapter will be the first to lay some of them to rest.

It all started in 1770 when Waring, in his book *Meditationes Algebraicae,* made the following conjecture: Every positive integer (he said in substance) can be expressed as the sum of no more than n positive kth powers, the value of n being a function of k. He stated that n equals 4 for squares, 9 for cubes and 19 for fourth powers, which is to say that every integer is the sum of no more than four squares or nine cubes or 19 fourth powers. For example, if you start investigating the natural numbers, you quickly discover that 7 is the smallest number that requires four squares $(4 + 1 + 1 + 1)$, but the next number, 8, requires only two $(4 + 4)$. Is there a number that requires more than four positive squares? Waring said no.

How about cubes? The smallest number requiring nine cubes is 23 (two 8's plus seven 1's). The smallest number requiring 19 fourth powers is 79 (four 16's and fifteen 1's), the smallest requiring 37 fifth powers is 223 (six 32's and thirty-one 1's) and so on. Waring's conjecture was that for every power there is a maximum number of positive powers required to express every natural number.

It seems unlikely that Waring, a mathematician of middling ability, had any powerful, secret techniques for proving his conjecture even for squares and cubes. More likely the values he gave for squares, cubes and fourth powers were just plausible surmises supported by flimsy empirical evidence.

At about the same time that Waring recorded his conjecture, his value of $n = 4$ for squares was verified. Fermat had believed this "four-square theorem" to be true, but Lagrange was the first to prove it. Some of the proofs are elementary but not short. Readers interested in working through a proof will find one clearly explained in *The Enjoyment of Mathematics*, by Hans Rademacher and Otto Toeplitz.

Hilbert's 1909 proof of Waring's general theorem was merely an existence proof; it gave no procedure for calculating the minimum number of kth powers. The proof, ingenious but difficult, was later improved in many ways. The simplest version, by the Russian number theorist Y. V. Linnik, is the last "pearl" in a marvelous little book by A. Y. Khinchin called *Three Pearls of Number Theory*. Linnik's proof is so "elementary," Khinchin says, that any good mathematician can master it in a mere three weeks of concentrated study!

As soon as mathematicians were convinced by Hilbert that Waring's theorem was true, they began searching in earnest for a formula that would give the minimum number of kth powers. The symbol $g(k)$, or "little gee" as it became known, was adopted as the symbol for this set of numbers. Euler had easily shown that the formula

$$\left[\left(\frac{3}{2}\right)^k\right] + 2^k - 2$$

where brackets indicate that the inside expression is rounded down to the nearest integer, gives a lower bound for $g(k)$. Mathematicians had long suspected that this formula is also an *exact* expression for $g(k)$. The suspicion was strengthened soon after Hilbert published his proof, when it was shown that $g(3)$ is indeed 9, as the formula predicts and as Waring had correctly guessed. That $g(4)$ equals 19 has not yet been proved. Leonard Eugene Dickson established 35 as a lower bound; then almost 40 years went by before modern computers allowed further progress. The value was lowered to 30 in 1971; the following year H. E. Thomas, Jr., brought it down to 23, where, so far as I know, it remains today. That $g(5)$ is 37 was proved in 1964 by Jing-jun Chen.

Since 1964 the value of "little gee" for $k = 6$ and all higher values has been shown to conform to Euler's formula for all k from 6 through 200,000. There are strong grounds for believing that no higher k values violate the formula. It has been known since 1957 that there is at most only a finite number of violations for k greater than 5, but the proof does not tell how to find such violations if indeed they exist. Most mathematicians today who have worked on Waring's problem are convinced that Euler's formula holds for all k, even though it has not been completely proved.

Much harder to pin down are the values of "big gee," or $G(k)$, the symbol for the smallest number of kth powers needed to express an infinite class of positive integers. In the case of squares the value of "big gee" was well known in Euler's day. It is 4, the same as the value of "little gee." It is not hard to show that integers of the form $4^a(8x + 7)$, where a and x are any nonnegative integers, require four squares to express them as a sum. The lowest such integer is 7, and the series continues: 15, 28, 23, 31, 39, ...

For cubes the situation is enormously more complicated; indeed, the value of $G(3)$ is far from known. Number theorists had long been puzzled by the fact that 23 and 239 were the only integers they could find that required as many as nine cubes to express them. (For example, 23, is the sum of the cubes of 2, 2, 1, 1, 1, 1, 1, 1, 1.) In 1939 Dickson proved the astonishing fact that among the infinity of integers only 23 and 239 require nine cubes. All integers above 239 can be expressed as the sum of eight or fewer cubes. Hence $G(3)$ is no higher than 8.

The value of $G(3)$ was soon lowered to 7 or less when it was established that only 15 integers require eight cubes: 15, 22, 50, 114, 167, 175, 186, 212, 231, 238, 303, 364, 420, 428 and 454. It is conjectured that beyond 454 all integers are the sums of seven or fewer cubes, but the conjecture remains far from proved.

"Big gee" for cubes is probably smaller than 7, but no one is sure. The largest number known to require seven cubes is 8,042 (the sum of the cubes of 19, 10, 4, 4, 3, 3, 1). It is known that there are infinitely many integers for which three cubes are not enough. The value of $G(3)$, therefore, is either 4, 5, 6 or 7. If the first figure is correct, as some number theorists hope, it means that there is a largest integer beyond which all integers can be expressed as the sum of no more than four positive cubes.

In view of the enormous difficulty of settling the value of $G(3)$, it is surprising that in 1939 the value of $G(4)$ was shown to be 16 by Harold Davenport. It is the only "big gee" above $G(2)$ that is known exactly.

An infinite number of integers can, of course, be expressed as the sum of one, two or three cubes. Do you remember the famous story about G. H. Hardy and his Indian friend Srinivasa Ramanujan? When Hardy visited Ramanujan in a hospital and told him that he had been taken there in a taxicab numbered 1729, an "uninteresting number," Ramanujan promptly responded that it was by no means uninteresting. It was the smallest integer that could be expressed as the sum of two cubes in two different ways. (1,729 is the sum of the cubes of 10 and 9, and 12 and 1.)

Many generalizations and variations of Waring's problem have been proposed, and the literature on such questions is voluminous. One of the oldest and

SQUARES	1		4		9		16		25		36 ...
FIRST DIFFERENCES		3		5		7		9		11 ...	
SECOND DIFFERENCES			2		2		2		2 ...		

Figure 135 Proof that eg(2) = 3 for "easier" Waring problem

most obvious variants is to allow negative powers as well as positive. The first important analysis of this problem was a 1934 paper by Hardy's associate E. M. Wright. (He and Hardy coauthored the classic *Introduction to the Theory of Numbers.*) Wright titled his paper "An Easier Waring's Problem," and it has been called that ever since in spite of the fact that finding values for $g(k)$ turned out to be incredibly difficult. Wright called it "easier" because it is easier to prove that $g(k)$ exists.

To avoid confusing the big and little "gees" of the easier Waring's problem from the classic one, let's put E or e (for "easier") in front of the g's. Of course, the existence of eg(k) follows immediately from Hilbert's 1909 proof, but Wright meant that, aside from the proof of Waring's problem, the existence of eg(k) when negative powers are allowed is much easier to obtain directly. But calculating eg(k) for the "easier" variant is quite a different matter. To this day it is known only for squares, in contrast to Waring's problem, where it is known for all values of k through 200,000 except for the stubborn case of fourth powers. "I had thought myself of writing a short expository article sometime on the 'easier' Waring problem," Wright said in a letter when I wrote to him for information, "to acknowledge quite how absurd my title for it has turned out to be."

The case for squares is trivial. An elementary application of the calculus of finite differences is sufficient to establish that eg(2) equals 3 for the easier problem. As shown in Figure 135, first put down a row of consecutive squares beginning with 1. The next row gives the differences between each adjacent pair of squares. Note that this row consists of consecutive odd numbers. It is clear that every odd number is the difference between two squares. Equally obvious is the fact that every even number can be expressed as the difference between two squares plus or minus 1. Since 1 is a square, we see that eg(2) equals 3. More formally, every number of the form $2x + 1$ (that is, every odd number) equals $(x + 1)^2 - x^2$, and every even number, $2x$, equals $x^2 - (x - 1)^2 + 1^2$, or $(x + 1)^2 - x^2 - 1$.

It is almost as easy to show that EG(2), "big gee" for squares, also is 3. Although some even numbers are themselves squares and others are the sum of, or difference between, two squares, there is an infinite class of even numbers (of the form $8x + 6$, where x is any nonnegative integer) that require three squares

	CUBES 1	8	27	64		125		216 ...
FIRST DIFFERENCES	7	19	37		61		91 ...	
SECOND DIFFERENCES		12	18	24		30 ...		
THIRD DIFFERENCES			6	6	6 ...			

Figure 136　Proof that eg(3) ≤ 5 for easier Waring problem

(positive or negative) for their expression. These are numbers in the series 6, 14, 22, 30, ...

It is hard to believe, but the values of "big gee" and "little gee" for cubes in the easier Waring problem and for all higher powers are still unknown. It is conjectured, but far from proved, that if negative powers are allowed, all positive integers can be expressed as the sum of no more than four cubes.

That five cubes suffice is easily demonstrated. Again we use the calculus of finite differences. The cubes provide the first row *[see Figure 136]*. Calculate two rows of differences. Note that numbers in the third row, 12, 18, 24, 30, 36, . . . , are consecutive multiples of 6. Each of these numbers can be expressed by four cubes. Consider 18. The table shows that 18 is the difference between the two numbers above it, 19 and 37. Nineteen is the difference between the two cubes above it, 8 and 27, and 37 is the difference between cubes 27 and 64. It follows that $18 = (64 - 27) - (27 - 8) = 4^3 - 3^3 - 3^3 + 2^3$. Clearly this procedure provides a four-cube expression for any multiple of 6.

It remains to show that numbers that are not multiples of 6 can be expressed by five cubes. Consider the five numbers between 18 and 24. They are 19, 20, 21, 22 and 23. Each of these numbers differs by a cube from a multiple of 6. The two end numbers, 19 and 23, differ by 1 from a multiple of 6, so that we can express 19 by adding 1 to the four cubes that express 18, and express 23 by subtracting 1 from the four cubes that express 24. The remaining numbers are 20, 21 and 22. Twenty differs by 2^3 from 12, and 22 differs by 2^3 from 30; therefore we can express 20 by adding 8 to the four cubes that express 12, and we can express 22 by subtracting 8 from the four cubes that express 24. Only 21 remains. It differs by 3^3 from 48; therefore we can express 21 by taking 27 from the four cubes that express 48. This procedure applies to all the numbers that lie between multiples of 6. By adding or subtracting a suitable cube to or from a set of four other cubes, we can express every nonmultiple of 6 by five cubes.

Hardy and Wright, in their *Introduction to the Theory of Numbers*, give a shorter proof, based on the fact that $n^3 - n = 0$ (mod 6). This leads to the following formula for expressing any number, n, with five cubes:

$$n = n^3 - 6x = n^3 - (x + 1)^3 - (x - 1)^3 + x^3 + x^3$$

where x is a suitable positive integer. Neither this formula nor the previous procedure indicates how to express n with the fewest number of cubes; they only show how to do it with five cubes. For instance, the procedure tells us that $15 = 8^3 - 7^3 - 7^3 + 6^3 - 3^3$. The Hardy and Wright formula leads to the more monstrous expression $15 = 15^3 - 561^3 - 559^3 + 560^3 + 560^3$. Actually 15 can be expressed with as few as three cubes: $2^3 + 2^3 - 1^3$.

The big question, still unresolved, is whether four cubes (allowing negative cubes) are sufficient for expressing every positive integer. No one has proved they are. No one has found a counterexample. The simplest expressions known to me, with four or fewer cubes, for integers 1 through 99 are listed in Figure 137. ("Simplest" means that the number of cubes is minimized, and an expression is given with the smallest absolute value for the cube of largest absolute value.) The chart is based on information supplied by George Shombert, Jr., of Beaver, Pa., supplemented by the results of two computer programs. In 1955 J. C. P. Miller and M. F. C. Woollett, using a computer at the University of Cambridge, searched for three-cube expressions of integers 1 through 100 within a range of cubes through $3,200^3$ (see the bibliography). In 1964 the search was extended through integer 999 within a range of cubes through $65,536^3$ by V. L. Gardiner, R. B. Lazarus and P. R. Stein, using a computer at the Los Alamos National Laboratory of the University of California (see the bibliography).

Note that four cubes are needed for every number with a digital root of 4 or 5. (The digital root of a number is obtained by adding the number's digits, then adding the digits in the result and continuing until one digit remains. It is the same as the value of the number modulo 9.) It is not hard to show that every number equal to 4 or 5 (mod 9) [shaded numbers in Figure 137] requires at least four cubes.

One way to do it makes use of the old accountant's trick for checking addition by digital roots. The sum of any set of integers, positive or negative, has a digital root equal to the root of the sum of the roots of the same set of numbers. Every cube has a digital root of 1, 8 or 9. No pair of these roots (the possible pairs are 11, 18, 19, 88, 89 or 99), whatever their signs, has a sum with a digital root of 4 or 5. (When adding a negative root, it is often convenient to change the minus sign to plus and the root to its complement with respect to 9. Thus $1 - 8$ is the same as $1 + 1$, giving the positive digital root of 2.) Consequently no integer equal to 4 or 5 (mod 9) can be expressed by two cubes.

Moreover, no triplet of digital roots 1, 8 or 9, whatever their signs, has a sum with digital root 4 or 5. Therefore no triplet of cubes can express a number equal to 4 or 5 (mod 9). We not only have established that eg(3) for the easier problem is at least 4 but also have found an infinite class of integers (those of form $9x + 4$ and $9x + 5$) that require at least four cubes. Both eg(3) and EG(3) are each at least 4.

$1 = 1^3$

$2 = 1^3 + 1^3$

$3 = 1^3 + 1^3 + 1^3$

$4 = 1^3 + 1^3 + 1^3 + 1^3$

$5 = 2^3 - 1^3 - 1^3 - 1^3$

$6 = 2^3 - 1^3 - 1^3$

$7 = 2^3 - 1^3$

$8 = 2^3$

$9 = 2^3 + 1^3$

$10 = 2^3 + 1^3 + 1^3$

$11 = 3^3 - 2^3 - 2^3$

$12 = -11^3 + 10^3 + 7^3$

$13 = -11^3 + 10^3 + 7^3 + 1^3$

$14 = 2^3 + 2^3 - 1^3 - 1^3$

$15 = 2^3 + 2^3 - 1^3$

$16 = 2^3 + 2^3$

$17 = 2^3 + 2^3 + 1^3$

$18 = 3^3 - 2^3 - 1^3$

$19 = 3^3 - 2^3$

$20 = 3^3 - 2^3 + 1^3$

$21 = 16^3 - 14^3 - 11^3$

$22 = 16^3 - 14^3 - 11^3 + 1^3$

$23 = 2^3 + 2^3 + 2^3 - 1^3$

$24 = 2^3 + 2^3 + 2^3$

$25 = 3^3 - 1^3 - 1^3$

$26 = 3^3 - 1^3$

$27 = 3^3$

$28 = 3^3 + 1^3$

$29 = 3^3 + 1^3 + 1^3$

$30 = 3^3 + 1^3 + 1^3 + 1^3$

$31 = 52^3 - 44^3 - 44^3 + 31^3$

$32 = 2^3 + 2^3 + 2^3 + 2^3$

$33 = 3^3 + 2^3 - 1^3 - 1^3$

$34 = 3^3 + 2^3 - 1^3$

$35 = 3^3 + 2^3$

$36 = 3^3 + 2^3 + 1^3$

$37 = 4^3 - 3^3$

$38 = 4^3 - 3^3 + 1^3$

$39 = -159{,}380^3 + 134{,}476^3 + 117{,}367^3$

$40 = 4^3 - 2^3 - 2^3 - 2^3$

$41 = 8^3 - 7^3 - 4^3 - 4^3$

$42 = 3^3 + 2^3 + 2^3 - 1^3$

$43 = 3^3 + 2^3 + 2^3$

$44 = 8^3 - 7^3 - 5^3$

$45 = 4^3 - 3^3 + 2^3$

$46 = 3^3 + 3^3 - 2^3$

$47 = -8^3 + 7^3 + 6^3$

$48 = 4^3 - 2^3 - 2^3$

$49 = 4^3 - 2^3 - 2^3 + 1^3$

$50 = -49^3 + 41^3 + 29^3 + 29^3$

$51 = -796^3 + 659^3 + 602^3$

$52 = 3^3 + 3^3 - 1^3 - 1^3$

$53 = 3^3 + 3^3 - 1^3$

$54 = 3^3 + 3^3$

$55 = 3^3 + 3^3 + 1^3$

$56 = 4^3 - 2^3$

$57 = 4^3 - 2^3 + 1^3$

$58 = 4^3 - 2^3 + 1^3 + 1^3$

$59 = 5^3 - 4^3 - 1^3 - 1^3$

$60 = 5^3 - 4^3 - 1^3$

$61 = 5^3 - 4^3$

$62 = 3^3 + 3^3 + 2^3$

$63 = 4^3 - 1^3$

$64 = 4^3$

$65 = 4^3 + 1^3$

$66 = 4^3 + 1^3 + 1^3$

$67 = 4^3 + 1^3 + 1^3 + 1^3$

$68 = 5^3 - 4^3 + 2^3 - 1^3$

$69 = 5^3 - 4^3 + 2^3$

$70 = -21^3 + 20^3 + 11^3$

$71 = 4^3 + 2^3 - 1^3$

$72 = 4^3 + 2^3$

$73 = 4^3 + 2^3 + 1^3$

$74 = 4^3 + 2^3 + 1^3 + 1^3$

$75 = 4^3 + 3^3 - 2^3 - 2^3$

$76 = -11^3 + 10^3 + 7^3 + 4^3$

$77 = 5^3 - 4^3 + 2^3 + 2^3$

$78 = -55^3 + 53^3 + 26^3$

$79 = 35^3 - 33^3 - 19^3$

$80 = 4^3 + 2^3 + 2^3$

$81 = 3^3 + 3^3 + 3^3$

$82 = 14^3 - 11^3 - 11^3$

$83 = 4^3 + 3^3 - 2^3$

$84 = 4^3 + 3^3 - 2^3 + 1^3$

$85 = 7^3 - 5^3 - 5^3 - 2^3$

$86 = -31^3 + 29^3 + 14^3 + 14^3$

$87 = 4{,}271^3 - 4{,}126^3 - 1{,}972^3$

$88 = 5^3 - 4^3 + 3^3$

Figure 137 Simplest cube expressions known for numbers 1 through 99

$89 = \quad -7^3 + \quad 6^3 + \quad 6^3$
$90 = \quad\quad 4^3 + \quad 3^3 - \quad 1^3$
$91 = \quad\quad 4^3 + \quad 3^3$
$92 = \quad\quad 4^3 + \quad 3^3 + \quad 1^3$
$93 = \quad\quad 7^3 - \quad 5^3 - \quad 5^3$
$94 = \quad\quad 7^3 - \quad 5^3 - \quad 5^3 + \quad 1^3$

$95 = \quad -22^3 + 20^3 + 14^3 - \quad 1^3$
$96 = \quad -22^3 + 20^3 + 14^3$
$97 = \quad\quad 5^3 - \quad 3^3 - \quad 1^3$
$98 = \quad\quad 5^3 - \quad 3^3$
$99 = \quad\quad 4^3 + \quad 3^3 + \quad 2^3$

We have learned still more. A check of the possible quadruplets of 1, 8 and 9 reveals just four patterns that give a digital root of 4. They are 1, 1, 1, 1; -8, 1, 1, 1; -8, -8, 1, 1; and -8, -8, -8, 1. Similarly, just four patterns give a digital root of 5: 8, 8, 8, 8; 8, -1, -1, -1; 8, 8, -1, -1; and 8, 8, 8, -1. Therefore, in searching for four-cube solutions of integers equal to 4 or 5 (mod 9), we need to consider only cubes with digital roots of 8 (cubes of 2, 5, 8, 11, . . .) or cubes with digital roots of 1 (cubes of 1, 4, 7, 10, . . .). All integers with roots *not* 4 or 5 have been shown to have expressions of four or fewer cubes, so only numbers equal to 4 or 5 (mod 9) remain in doubt.

Observe that most of the numbers on the chart that are not equal to 4 or 5 (mod 9) have known three-cube expressions. Some were not easy to come by, notably the expression for 87 in which each cube has four digits. Miller and Woollett failed to find this expression, but it was trapped by the Los Alamos program.

The number 100 has an elegant four-cube expression: It is the sum of the cubes of 1, 2, 3 and 4. However, three three-cube expressions of 100 are known. In the simplest, each cube is a digit. Can the reader find this expression before checking the answer section where I shall give all three?

Many number theorists believe, although it has not been proved, that all integers not equal to 4 or 5 (mod 9) have three-cube expressions. If so, the value of eg(3) for the easier Waring problem is settled. Do you see why? To obtain a four-cube expression for a number equal to 4 (mod 9), we have only to add 1^3 to a three-cube expression for the number just below it; to obtain a four-cube expression for a number equal to 5 (mod 9), we have only to subtract 1^3 from a three-cube expression for the number just above it. Perhaps readers can discover three-cube solutions for 30, 33, 42, 52, 74, 75 and 84, which may have three-cube expressions, none of which has yet been found. In particular, can anyone express 30 with three cubes or prove that it cannot be done? It is astonishing that this continues to be an open question.

The number 12 is also of special interest. It is the smallest integer known for which only one three-cube solution is known. Most integers with three-cube expressions can be expressed by three cubes in more than one way. In some cases a number can be expressed by three cubes in infinitely many ways. For example, 2 can be expressed by substituting any positive integer for x in the following identity: $(6x^3 + 1)^3 - (6x^3 - 1)^3 - (6x^2)^3$. On the other hand, no one has yet found more than two ways to express 3 (cubes of 1, 1, 1 and -5, 4, 4). There is a noticeable paucity of three-cube solutions for numbers with digital roots of 3 and 6. Five of the numbers listed above for which three-cube solutions have yet to be found are numbers of this form.

Number theorists distinguish between primitive and derived solutions. A primitive solution is one in which the cubes have no common factor. A derived solution is one in which they do. Derived solutions are obtainable from primitive solutions by multiplying each cube root by n and the number itself by n^3, where n is any integer greater than 1. Twelve solutions in Figure 137 are derived: 16, 24, 32, 40, 48, 56, 64, 72, 80 and 96 (by multiplying primitive solutions by 2), and 54 and 81 (by multiplying primitive solutions by 3). For example, the solution for 16 is obtained from the primitive solution for 2 by multiplying 2 by 2^3 and each cube root by 2. The solution for 54 is derived from the primitive solution for 2 by multiplying 2 by 3^3 and each cube root by 3. The derived solutions for 24 and 80 are the only known solutions for those numbers. Note that a derived solution is not always the simplest. From $11 = 3^3 - 2^3 - 2^3$ we can derive $88 = 6^3 - 4^3 - 4^3$, but the primitive solution on the chart is simpler.

What about eg(4), the smallest number of fourth powers needed to express any integer? In 1941 W. Hunter (see the bibliography) showed that the number is either 9 or 10. For powers higher than four, the spread between upper and lower bounds is much greater.

ANSWERS

The problem was to express 100 as the sum of three cubes, allowing each cube to be positive or negative. The three known solutions are $7^3 - 6^3 - 3^3$, $190^3 - 161^3 - 139^3$ and $1{,}870^3 - 1{,}797^3 - 903^3$.

BIBLIOGRAPHY

Waring's Problem

"Waring's Problem and Related Results." Leonard Eugene Dickson in *History of the Theory of Numbers*, Vol. 2, Chapter 25. Chelsea, 1952. Reprint of the 1919 edition.

"An Elementary Solution of Waring's Problem." A. Y. Khinchin in *Three Pearls of Number Theory*, Chapter 3. Graylock, 1952.

"On Waring's Problem." Hans Rademacher and Otto Toeplitz in *The Enjoyment of Mathematics*, Chapter 9. Princeton University Press, 1957.

"Waring's Problem." W. J. Ellison in *American Mathematical Monthly*, Vol. 78, 1971, pages 10–36. Lists 146 references.

"Waring's Problem." I. N. Herstein and I. Kaplansky in *Matters Mathematical*, Chapter 2, Section 8. Harper and Row, 1974.

"Waring's Problem." Charles Small in *Mathematics Magazine*, Vol. 50, 1977, pages 12–16.

The Easier Waring's Problem

"An Easier Waring's Problem." E. M. Wright in *The Journal of the London Mathematical Society*, Vol. 9, 1934, pages 267–272.

"The 'Easier' Waring Problem." W. H. J. Fuchs and E. M. Wright in *The Quarterly Journal of Mathematics*, Vol. 10, 1939, pages 190–209.

"The Representation of Numbers by Sums of Fourth Powers." W. Hunter in *The Journal of the London Mathematical Society*, Vol. 16, 1941, pages 177–179.

"Solutions of the Diophantine Equation $x^3 + y^3 + 2^3 = k$." J. C. P. Miller and M. F. C. Woollett in *The Journal of the London Mathematical Society*, Vol. 30, 1955, pages 101–110.

"Solutions of the Diophantine Equation $x^3 + y^3 = z^3 - d$." V. L. Gardiner, R. B. Lazarus and P. R. Stein in *Mathematics of Computation*, Vol. 18, 1964, pages 408–413.

Tables of Solutions of the Diophantine Equation $x^3 + y^3 = z^3 - d$. V. L. Gardiner, R. B. Lazarus and P. R. Stein. Los Alamos Scientific Laboratory informal report UC-32, issued November 1973. Recent copies (Number LA-UR-85-2540) can be obtained from Paul Stein, Los Alamos National Laboratory, POB 1663, Los Alamos, NM 87545.

CHAPTER NINETEEN

Cram, Bynum and Quadraphage

There are many simple two-person games (for example, nim) for which perfect-play strategies are known. Other games, such as ticktacktoe and dots-and-squares, may appear just as simple but actually are so complex that no strategy has yet been found except when the game is played on special fields. In this chapter we consider several elegant new games that have extremely simple rules and about which relatively little is known. Some may not have general strategies; if they do, perhaps a reader of this book will be the first to discover them.

Our first game, as far as I am aware, has not been described before in print, although a few mathematicians have been involved with it since the early 1950's. I originally heard of it from Geoffrey Mott-Smith, the author of several books on games and puzzles. He told me it had been invented by a friend, who called it "plugg." Since then I have received letters from a number of mathematicians who independently invented the same game. In 1966 John Horton Conway gave it some thought, and although he did not succeed in cracking it, he did formulate a partial strategy with which the final stages of a game could be analyzed by standard nim theory.

The game can be played in various ways, all isomorphic. If the "board" is a rectangular lattice of dots in unit square formation, the rules are as follows. Players alternately draw a line that connects two orthogonally adjacent dots. No line may touch a dot after it has been joined to another. In the standard form of the game the last player to connect two dots is the winner. (In *misère,* or reverse, play the last to move is the loser.) Let's call the game "dots-and-pairs." Clearly it is a graph-theory game.

If a supply of counters is handy, one can arrange the counters to form the lattice; then each move consists in removing two orthogonally adjacent counters. Another way to play dots-and-pairs is to sketch a checkered field on

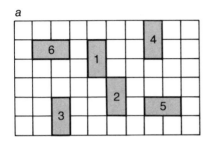
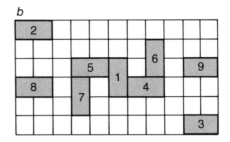

Figure 138 Symmetrical winning strategy for cram on even-even fields *(a)* and even-odd fields *(b)*

paper or outline it on graph paper. A move consists in coloring two orthogonally adjacent cells or, more simply, drawing a line that eliminates the "domino."

Still another way to play the game — the most pleasant — is to place dominoes on a checkerboard. The markings on the dominoes are, of course, irrelevant. All that matters is that each domino must cover just two squares. Players alternately place one domino until no further play is possible. We shall call this version of the game "cram." It is the simplest nontrivial polyomino-placing game.

Winning strategies for cram are known only for certain boards. If, for instance, the field is rectangular with two even sides and the game is standard (the last to play wins), the second player wins easily by symmetry play. He simply makes each move symmetrically opposite his opponent's last move *[see Figure 138a]*. To eliminate this strategy, we can add a new rule: The second player's first move must not be symmetrical with respect to the first player's opening move. With this proviso the game can be played on a standard chessboard with 32 dominoes. It is not known which player has the win.

If standard cram is played on an even-by-odd rectangle, the first player wins by taking the two central cells, then playing symmetrically thereafter *[see Figure 138b]*. This strategy can be eliminated by denying the center to the first player on his opening move.

No general strategy is known for the reverse form of cram on even-even or even-odd fields, and no general strategy is known for standard or reverse play on odd-odd fields. Even when one of the odd sides in an odd-odd field is reduced to 1, the game is complex and still unsolved. In 1973 David Singmaster, then at the Istituto Matematico in Pisa, wrote a computer program for the 1-by-m field, standard game, with m less than 1,000. Assuming that the first player loses (because he cannot move) when m equals 0 or 1, Singmaster found 151 values of m that give the win to the second player. For m less than 100 the values are 0, 1, 5, 9, 15, 21, 25, 29, 35, 39, 43, 55, 59, 63, 73, 77, 89, 93, 97.

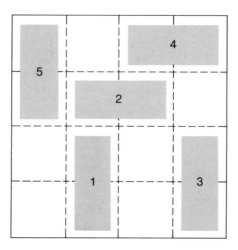

Figure 139   Standard crosscram with a first-player win

When m is even, the first player wins, of course, by taking the center and playing symmetrically. When m is odd, he wins for all values less than 100 that are not in the above set. I know of no computer analysis of the reverse game for 1-by-m fields.

Square fields of only the lowest orders have been investigated. The 3-by-3 game is trivial. It takes just a few minutes of analysis to see that the second player wins the standard game and loses the reverse game. The 4-by-4, with symmetry play denied the second player in standard play, takes considerably more work. Conway found it to be a second-player win in both standard and reverse play.

What about the 5-by-5? Because it is odd-by-odd, symmetry strategy is ruled out and no special rules are needed. Who has the win in standard play? In *misère?* As far as I know, neither question has yet been answered.

Cram is an "impartial" game. This means that any possible move can be made by either player. "Partial" games, or what Conway prefers to call "unimpartial" games, are those in which moves open to one player are denied to the other. Chess and checkers, for example, are partial games because each player can move only the pieces of his color. We can convert cram to a partial game by a rule proposed by Göran Andersson, who wrote to me about it in 1973.

The rule is delightfully simple. One player must make all his moves horizontally; the other must make all his moves vertically. Call it "crosscram." This, of course, instantly eliminates symmetry play from all rectangular boards. The 3-by-3, as before, can be quickly disposed of. The first player wins the standard game provided that his first move does not include a corner cell. The second player wins the reverse game. Crosscram in the 4-by-4 form is sufficiently complicated to make a good pencil-and-paper game *[see Figure 139]*. Does the first or second player win the standard game? The reverse game?

Both cram and crosscram can be regarded as special cases of more general games. Cram is a special case of Piet Hein's Tac Tix, now more commonly

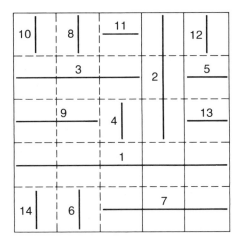

Figure 140   Bynum's game with a second-player win

called nimbi. (See Chapter 15 of *The Scientific American Book of Mathematical Puzzles & Diversions.*) Nimbi is usually played with a square array of counters. A move consists in removing as many orthogonally adjacent counters as desired, not just a pair, from any orthogonal row. Crosscram is a special case of partial nimbi in which the rules restrict one player to horizontal rows and the other to vertical.

Another interesting form of partial nimbi was invented in 1972 by James Bynum of Tacoma, Wash., who has permitted me to describe it here. It is the same as partial nimbi except that each play must have a maximum length; that is, the orthogonally adjacent cells must be bounded on each end by either the field's border or an opponent's perpendicular move. The game's first move must necessarily be an entire row or column *[see Figure 140]*.

Bynum's game was solved in 1973 by Conway. *Misère* play is almost trivial. The second player wins on all square boards, and if the board is a nonsquare rectangle, the player whose moves parallel the shorter dimension wins regardless of whether he goes first or second. The winning rule is this: Pick one of the two sides that parallel your moves, then always play as close to that side as possible. Standard play is more interesting. The first player has the win on all square boards as well as on all nonsquare rectangles with sides of the same parity (even-even or odd-odd). On even-odd rectangles the player whose moves parallel the even dimension wins regardless of who plays first.

Interested readers may enjoy seeing if they can develop a set of strategic rules that will ensure a win for the player who has the win. The game yielded readily to Conway's theory of unimpartial games. I shall say no more about his analysis because it will be included in a book on unimpartial game theory that Conway is reported to be writing.

Quadraphage (square-eater) is a partly explored family of games invented and named by David L. Silverman in the late 1940's. He suggested the basic idea to Richard A. Epstein, who mentions one version briefly on page 406 of his *The Theory of Gambling and Statistical Logic*. Silverman discusses two other versions on page 186 in his book of game puzzles, *Your Move*. Elwyn Berlekamp has done considerable work on quadraphage, which he will summarize in a book on games that he is preparing in collaboration with Conway and Richard K. Guy. Here I shall introduce only some of the game's simpler aspects.

Quadraphage games are played on a chessboard of side n, usually square. Pieces consist of one chess piece, usually a king, and a supply of counters. The counters are the quadraphages, which I shall call quads for short. Each quad "eats" the cell on which it is placed, thus preventing the king from moving to that cell. The game starts with the king on the central square if the board is odd-odd or on one of the four central squares if the board is even-even. The rest of the board is empty. One player moves the king in the usual manner. The other player places counters, q at a time, on Q cells. As in the game of go, counters do not move once they are placed. The object of the king is to get safely to the edge of the board. The quads try to box in the king so that it cannot escape. The quad player conventionally goes first; then the players move alternately.

If q equals 4 (four squares eaten on each move), it is easy to show that the king can be captured in no more than three moves on all boards of side 5 or greater. (Of course, the king escapes immediately on a 4-board.) If q equals 3, it takes only a little more effort to find that the king can be trapped on boards of side 6 or greater.

When q equals 2, the game starts to get interesting. The king escapes on the 7-board but can be trapped on the 8-by-8 and all larger boards. The strategy is to move first as shown in Figure 141. If the king's first move is toward one of the white corner cells, say the northwest corner, one quad goes on the cell two squares west of the king, and the other quad goes on the cell immediately north of that square. Thereafter (and also if the king makes any other first move except toward a corner white cell) the strategy is to block the king by placing quads on white border cells. When the king is adjacent to the border, then, of course, quads must go side by side to prevent his escape.

What about $q = 1$? Can the king always escape no matter how large the board? Surprisingly, it cannot. Berlekamp has proved that on a board as small as 33-by-33 the king is lost. Unfortunately both proof and strategy are too complicated to give here. Golomb has shown that if the king's moves are limited to orthogonal moves and q equals 1, the king escapes on the 7-board but can be trapped on the 8-board.

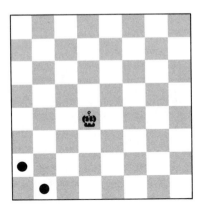

Figure 141 Two quads to trap king on an 8-by-8 board

Although the king escapes easily on a standard chessboard, a pleasant game (proposed by Silverman) results if the king tries to maximize its moves before reaching the border. Quads, placed one per move, try either to trap the king or to force it to the border as quickly as possible. If the king escapes, it is awarded one point for each quad on the board. No points are awarded if the king is trapped. The game is particularly enjoyable when played on the order-18 go board.

The go board, with its large supply of stones, is a handy tool for working on the many unsolved quadraphage problems. What, for instance, happens when the king is replaced by a different chess piece? If the piece is a bishop, rook or queen, we must limit the length of its move to avoid triviality. Assume that the board is infinite but the piece cannot move a distance of more than, say, a billion squares. With this limitation a bishop is easily captured in a goose chase along a diagonal by placing three quads per move as shown in Figure 142a. Once the ends are blocked, the bishop can be confined to the diagonal. Three quads per move can similarly trap a rook in a goose chase along an orthogonal, as shown in *b*, and seven can trap a queen along either an orthogonal or a diagonal. Can the bishop or rook be trapped by two quads per move? By one? Can the queen be trapped by fewer than seven?

If the piece is a knight, we must consider it free when it lands on a cell that is second from the border, because on its next move it can leap over a border cell. On boards of sides 5, 6 and 7 the centered knight has eight moves to freedom, so eight quads obviously are needed to trap it. Five quads per move will trap the knight on the order-8 board, and four per move are sufficient for the order-9. Figure 143 shows one of several winning first moves for each board, although I must add that I am not certain all this information is correct.

Can three quads per move trap a knight on an infinite board? One quad per

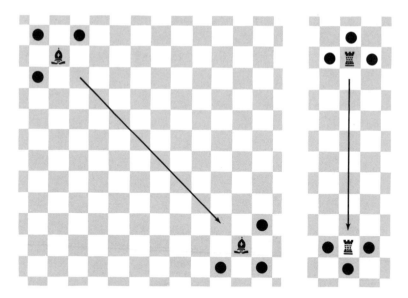

Figure 142 How three quads per move trap bishop *(a)* and rook *(b)* on infinite boards

move surely is insufficient, although Berlekamp reports he has yet to see a rigorous proof.

ANSWERS

The first player wins the direct 4-by-4 crosscram game, but only if he takes two central cells or two at the middle of a side. The first player also wins the reverse game. There seem to be no simple strategies for either form of play, and the game trees are too complicated to give here. Several readers sent proofs that the second player wins 5-by-5 direct crosscram but loses the reverse game.

Although crosscram remains unsolved in general for both direct and reverse

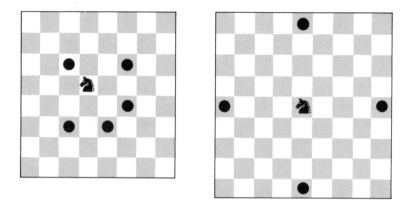

Figure 143 Trapping a knight on order-8 board (left) and order-9 board (right)

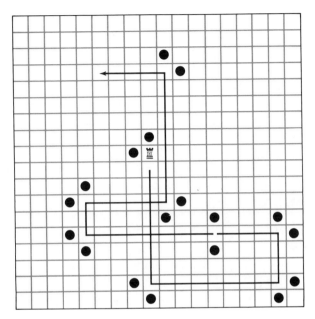

Figure 144   How to trap a rook with two quads per move

play, Bynum's game is solved for all rectangular boards. You'll find a complete analysis, along with several delightful variations, in Chapter 15 of Conway's *On Numbers and Games*. Conway calls Bynum's game "one of the most interesting we have studied." More on the game, as well as discussions of crosscram, which the authors call the "domineering" game, can be found in *Winning Ways* (see the bibliography).

If a rook's maximum move in a game of quadraphage is n cells, it can be trapped by two quads per move on a board of side $2n + 2$. The strategy is to consider the unobstructed paths from the rook to sides of the board and to place the quads adjacent to the rook to block its movement to the two nearest sides. (If two borders are the same distance away, choose either one.) Figure 144 shows the strategy on a go board, the rook limited to maximum moves of eight cells. The rook clearly can never reach the edge. Eventually it must head toward a quad. When this happens, quads on each side confine it to a segment of the path, and entrapment quickly follows.

The same procedure will trap a bishop within the borders of a sawtooth board that is $2n + 2$ on the side, where n is the longest move allowed for the bishop. Figure 145 shows how the strategy operates on a 18-by-18 sawtooth board when the bishop can move no more than eight cells along the diagonals. Such boards of $2n + 2$ on the side are (for bishops) isomorphic with chessboards of $4n + 3$ on the side. The sawtooth board shown is therefore equivalent

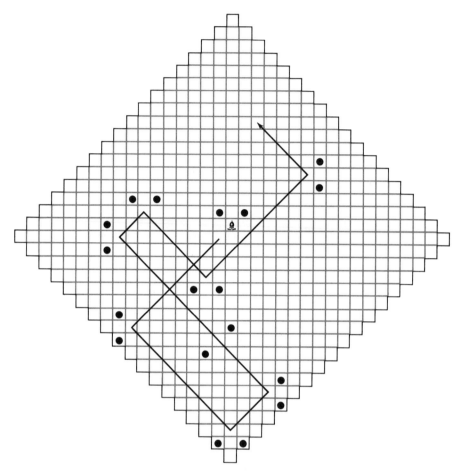

Figure 145   How to trap a bishop with two quads per move

to a 35-by-35 chessboard on which the bishop is confined to cells that have the same color as the corner squares.

A similar strategy using four quads per move will trap the queen. For *n* greater than 2 a board of at least $4n + 2$ on a side seems necessary. (Thus a queen with maximum move 8 can be trapped on a 34-sided field.) The first move is to put quads on the four corner cells. Thereafter use the "nearest sides" strategy. When there is a choice between blocking equal paths on an orthogonal and a diagonal, block the orthogonal.

ADDENDUM

Many readers resolved the question of who wins 5-by-5 cram. The second player wins the direct game but loses when the play is reversed. A correspondent in Sweden, Magnus Tideman, sent the most complete analysis. He also showed that the first player wins the direct 3-by-5 game but loses the reverse

3-by-5. The second player wins the reverse 4-by-5. The first player wins reverse 3-by-6. On the 3-by-7 field the first player wins the direct game but loses when played in reverse. No general strategies are known.

A variation of cram, played by placing dominoes on a 6-by-6 field according to special rules, was on sale in France in the early 1970's under the trade name Cogito.

When cram is played on a 1-by-m field—let's call it linear cram—it is a game that was proposed as early as 1934 by T. R. Dawson, writing in the December issue of his *Fairy Chess Review*. He described a generalization of kayles in which players alternately remove k adjacent counters from n rows. If $n = 1$ and $k = 2$, the game is linear cram. The game is also isomorphic with Regulus, presented as an unsolved game in David L. Silverman's book *Your Move*.

In its standard (direct) form, linear cram was completely solved in 1956 by Richard K. Guy and Cedric A. B. Smith in their classic paper on nimlike games, "The G-values of Various Games" (see the bibliography). In their notation the game has the name of .07 for reasons we need not go into here. As noted earlier, when m is even, the first player wins by taking the center and playing symmetrically. When m is odd, the game develops a curious periodicity of 34. The complete solution is that the second player wins on all boards, and only those boards, with a number of cells equal to 0, 1, 15, 35 or (modulo 34) to 5, 9, 21, 25 and 29.

In reverse form, linear cram remains unsolved. Computer programs for fields up through 43 in length show second-player wins for $m = 2, 3, 7, 8, 12, 16, 17, 21, 22, 26, 30, 31, 35, 36$ and 40. There is a periodicity here of 14—the figures conforming to 2, 3, 7, 8, 12 (modulo 14)—but whether this periodicity persists for higher values of m is still not known. Ashok Chandra, an IBM research mathematician, found an elegant proof that the second player can win reverse cram on all fields of 2-by-$(2m + 1)$.

When linear cram is played with straight trominoes (1-by-3 rectangles), Guy and Smith call it the .007 game. This "James Bond" version, as Conway refers to it in *Winning Ways*, is solved only for the direct game on odd fields. The first player wins by taking the center and playing symmetrically. When m is even, no periodicity has yet been detected in the game's G-values (Grundy numbers), nor has tromino cram been solved for either odd or even m in the reverse form of play. Conway has shown that tromino cram is isomorphic with ticktacktoe played on a linear field when the object is to get three adjacent marks and both players use the *same* mark. Frank Harary, the graph theorist, calls it one-color linear ticktacktoe. For such a simple game, it is astonishing that it is so difficult to solve.

In the game of quadraphage, many readers sent proofs that one quad per move is sufficient to trap a rook or bishop, and three quads per move will catch a queen. In the rook case, assume the rook is limited to n cells per move. The trapping strategy, on the minimum board of side $8n^2 + 3$, is to use the first $4n$ moves to place (regardless of how the rook moves) n quads at the top of each top corner and at the bottom of each bottom corner. All four corners can be sealed in this way before the rook can attack a corner cell. Since the rook can then attack only one border cell at a time, single quads suffice to complete the entrapment.

As explained earlier, a bishop on a sawtooth board is equivalent to a rook on a regular board. For that reason the strategy just described will trap a bishop with one quad per move on a sawtooth-bordered board of side $8n^2 + 3$ or the isomorphic chessboard of side $16n^2 + 5$. A similar strategy traps a queen with three quads per move on a board of side $2n[8n/3] + 3$, where n is the maximum distance the queen is permitted to move and the brackets indicate rounding up to the nearest integer. Regardless of the queen's initial moves, the strategy is to place $2n$ quads (three per move) on both sides of each corner. This prevents the queen from attacking more than three border cells on all subsequent moves. Chandra showed that if the queen's maximum move is 2 it can be trapped by two quads per move on a board of side 67 or possibly smaller.

Several readers found that the knight can be trapped with three quads per move. E. N. Adams showed how to do this on the 19-by-19 go board and possibly on a board as small as 16-by-16. A surprising letter from Jerry Butters gave a procedure for trapping a knight with just two quads per move. His 13-page proof requires a board of side 4,500. This size can surely be reduced considerably, but at present there are only conjectures about how small the board can be. If only one quad is placed at a time, the knight can probably escape on the infinite board, but no one seems close to proving it.

Vast regions of quadraphage remain unexplored. We can, for example, raise questions about the trapping of two or more chess pieces not necessarily alike. We can consider "fairy" chessmen, such as a superqueen that also moves like a knight, or a rook or bishop with the added knight move. Since we can invent bizarre chessmen that move according to any specified set of rules, the range of quadraphage-type games obviously is unlimited.

Conway has considered such pieces as the Angel, which has the power to move to any cell that can be reached by n king moves, with n being given any value you like. For example, n can be 1,000. Because the Angel has wings, it can fly over quads to any vacant cell within its 1,000 range. As the authors of *Winning Ways* put it, the Devil (who plays the quads) "wins if he can surround the Angel with a sulphurous moat, a thousand squares wide, of eaten squares." With the Devil limited to one quad per move, it seems probable that the angel can always escape, although no one has found an explicit strategy to prove it. In addition to Conway, Andreas Blass, a mathematician at the University of Michigan, has given some thought to the Angel and to devilish tactics for trapping it. Blass and Conway have proved the surprising result that the Devil, limited to only one quad per move, can infinitely often compel the Angel to move inward for any arbitrary finite distance! The proof involves constructing arcs of quads that force the Angel to back up in order to get around a blocking arc.

Blass and Conway have also studied such chessmen as the Pope, an angel without wings; the Fool, an angel that never moves south; and the Flee, an angel that is not allowed to move inward toward the cell where it started. Blass has shown that the Fool, which rushes in where Angels fear to tread, can always be trapped on a sufficiently large board by the Devil playing one quad per move.

You'll find a brief discussion of quadraphage in *Winning Ways,* volume 2, along with an extension of the game in which the square-eater places both black and white go stones on each move. The black stones remain fixed, but a white stone can be moved to any empty cell, leaving uneaten the cell from which it moved. At each move the square-eater has three options:

1. Place one stone of either color on a vacant cell.

2. Move a wandering (white) stone.

3. Pass.

BIBLIOGRAPHY

"The *G*-values of Various Games." R. K. Guy and C. A. B. Smith in *Proceedings of the Cambridge Philosophical Society,* Vol. 52, Part 3, 1956, pages 514–526.

The Theory of Gambling and Statistical Logic. Richard A. Epstein. Academic, 1967.

Your Move. David L. Silverman. McGraw-Hill, 1971.

On Numbers and Games. J. H. Conway. Academic, 1976.

"Reverse Cram with Block Sizes Not Exceeding 13." Ronald Evans in *Delta,* Vol. 6, 1976, pages 57–66.

Winning Ways. Elwyn R. Berlekamp, John H. Conway and Richard K. Guy. Academic, 1982.

CHAPTER TWENTY

The I Ching

The *I Ching* (pronounced *ee jing*), or *Book of Changes,* is one of the world's oldest books and also one of the most enigmatic. For more than 2,000 years it has been used in the East as a book of divination, and it still is studied with awesome reverence as a rich source of Confucian and Taoist wisdom. Tens of thousands of young people in the U.S. (particularly in California), caught up in the current occult explosion and eager to know more about Eastern mysticism and early Chinese history, are now consulting the *I Ching* as seriously as they consult the Ouija board or the tarot cards. C. G. Jung was convinced of the *I Ching*'s extraordinary power to foretell the future; he even asked it about the prospects of American sales of a new English translation of itself and got an optimistic answer. More recent pundits who are deep into occultism — England's Colin Wilson, for example — have written about their experiences with the *I Ching*'s terrifying oracular accuracy.

The early history of the *I Ching* is unknown. Most likely it began as early as the eighth century B.C. as a collection of peasant omen-texts; then slowly over the centuries these documents became combined with stick divination practices. A few centuries before Christ, near the end of the Chou dynasty, it acquired its present form and became one of the five great classics of the Confucian canon.

The combinational foundation of the *I Ching* consists of 64 hexagrams. They display every possible permutation of two types of line when taken six at a time. Each hexagram has a traditional Chinese name. The two kinds of line proclaim the basic duality of Chinese metaphysics: The broken line corresponds to yin, the unbroken line to yang. If we take the lines two at a time, there are $2^2 = 4$ ways to combine them into what are called digrams, and $2^3 = 8$ ways to form trigrams. The trigrams, with their Chinese names and symbolic meanings, are shown in Figure 146.

TRIGRAM	NAME	IMAGES	TRAITS	FAMILY RELATIONS	PARTS OF BODY	ANIMALS
☰	CH'IEN	HEAVEN COLD	STRONG FIRM LIGHT	FATHER	HEAD	HORSE
☷	K'UN	EARTH HEAT	WEAK YIELDING DARK	MOTHER	BELLY	OX
☳	CHÊN	THUNDER SPRING	ACTIVE MOVING AROUSING	FIRST SON	FOOT	DRAGON
☵	K'AN	WATER MOON WINTER	DANGEROUS DIFFICULT ENVELOPING	SECOND SON	EAR	PIG
☶	KÊN	MOUNTAIN	RESTING STUBBORN UNMOVING	YOUNGEST SON	HAND	DOG
☴	SUN	WIND WOOD	GENTLE PENETRATING FLEXIBLE	FIRST DAUGHTER	THIGH	BIRD
☲	LI	FIRE SUN LIGHTNING SUMMER	BEAUTIFUL DEPENDING CLINGING	SECOND DAUGHTER	EYE	PHEASANT
☱	TUI	LAKE MARSH RAIN AUTUMN	JOYFUL SATISFIED COMPLACENT	YOUNGEST DAUGHTER	MOUTH	SHEEP

Figure 146   The eight trigrams and some of their meanings

There are two ancient ways of displaying the eight trigrams in a circle. The oldest, known as the Fu Hsi arrangement after the mythical founder of China's first dynasty (the Hsia dynasty, 2205 – 1766 B.C.) is shown at the left in Figure 147. Note that opposite pairs are complementary both in symbolic meaning and in the mathematical sense that each is obtained from the other by replacing yin lines with yang and yang lines with yin. This arrangement, usually surrounding the familiar yin-yang symbol, is still widely used throughout China,

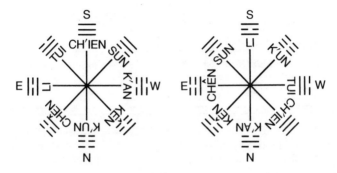

Figure 147   Fu Hsi arrangement of trigrams (left) and King Wen arrangement (right)

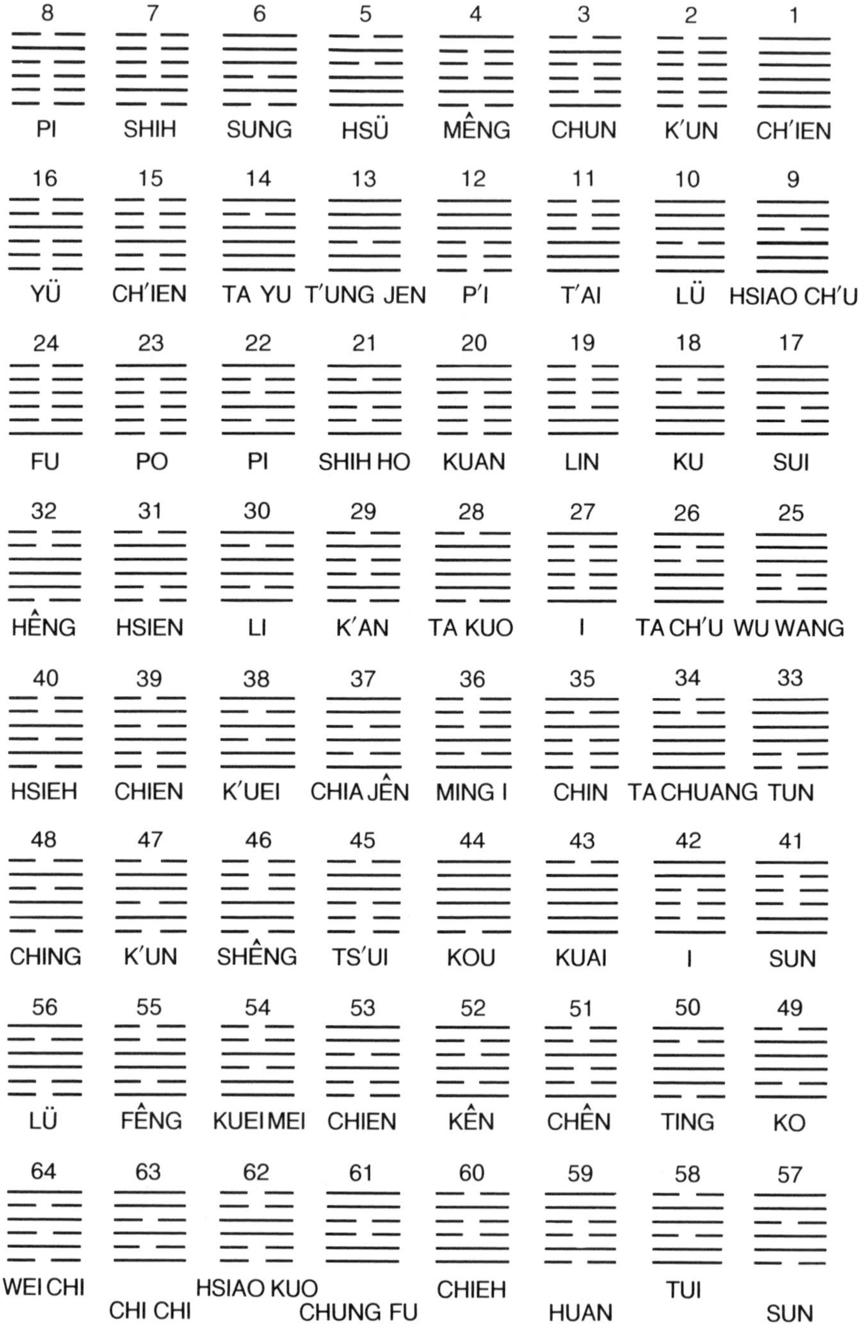

8	7	6	5	4	3	2	1
PI	SHIH	SUNG	HSÜ	MÊNG	CHUN	K'UN	CH'IEN

16	15	14	13	12	11	10	9
YÜ	CH'IEN	TA YU	T'UNG JEN	P'I	T'AI	LÜ	HSIAO CH'U

24	23	22	21	20	19	18	17
FU	PO	PI	SHIH HO	KUAN	LIN	KU	SUI

32	31	30	29	28	27	26	25
HÊNG	HSIEN	LI	K'AN	TA KUO	I	TA CH'U	WU WANG

40	39	38	37	36	35	34	33
HSIEH	CHIEN	K'UEI	CHIA JÊN	MING I	CHIN	TA CHUANG	TUN

48	47	46	45	44	43	42	41
CHING	K'UN	SHÊNG	TS'UI	KOU	KUAI	I	SUN

56	55	54	53	52	51	50	49
LÜ	FÊNG	KUEI MEI	CHIEN	KÊN	CHÊN	TING	KO

64	63	62	61	60	59	58	57
WEI CHI	CHI CHI	HSIAO KUO / CHUNG FU	CHIEH	HUAN		TUI	SUN

Figure 148 King Wen arrangement of the 64 I Ching hexagrams

Japan and Korea as a good-luck charm to put over doorways and on jewelry. It is also called the "earlier heaven" or "primal arrangement." The King Wen arrangement (after the legendary father of the founder of the Chou dynasty), shown at the right in Figure 147 (also called the "later heaven" or "inner-world arrangement"), abandons the complementary positioning of the Fu Hsi sequence, so the trigrams at the cardinal points of the compass symbolize the seasons in cyclic order. If one starts at south (traditionally shown at the top) and going clockwise, the hexagrams at the cardinal points stand for summer, fall, winter and spring.

The oldest way of arranging the 64 hexagrams, which is known as the King Wen sequence, is the order in which they appear in the *I Ching [see Figure 148]*. The rows are taken from right to left as the numbering indicates. Note that the hexagrams are paired in a singular way. Each odd-numbered hexagram is followed by a hexagram that is either its inverse or its complement. If the odd hexagram has twofold symmetry (is the same upside down), it is complemented to produce the next hexagram. If it lacks twofold symmetry, it is inverted.

Is there any kind of mathematical order that determines the sequence in which the hexagram pairs follow one another? This is an unsolved problem. From time to time a student of the *I Ching* announces his discovery of a mathematical scheme underlying the arrangement of pairs, but on closer inspection it turns out that so many arbitrary assumptions are made that, in effect, the order must be assumed before it emerges from the analysis. As far as anyone knows, the pairs of the King Wen sequence are in random order, and there is no known basis for determining which member of a pair precedes the other.

Not until the 11th century did Chinese scholars discover a very simple and elegant way to order the hexagrams. This arrangement is attributed to Fu Hsi *[see Figure 149]*. The white space at the bottom represents the *t'ai chi,* the state of

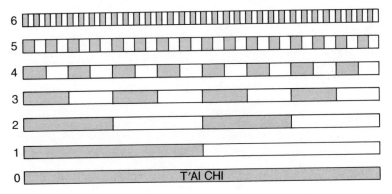

Figure 149   How six yin-yang divisions generate the 64 hexagrams

the universe when it was "without form, and void" (as Genesis 1:2 puts it). This undifferentiated chaos divides into the yin (gray) and yang (black) halves of the row labeled 1. In row 2 we see the yin dividing into yin and yang, and similarly the yang. These binary divisions continue upward for six steps.

The chart now automatically gives all the polygrams of orders 1 through 6. Divide rows 1 and 2 vertically into four equal parts, replace the gray in each part with broken (yin) lines and you have the four digrams. Rows 1, 2 and 3, divided vertically into eight equal parts, give the eight trigrams. Rows 1, 2, 3 and 4, in 16 parts, give the 16 tetragrams, rows 1, 2, 3, 4 and 5, in 32 parts, give the 32 pentagrams, and rows 1, 2, 3, 4, 5 and 6, in 64 parts, give the 64 hexagrams. Figure 150 shows the hexagrams in their traditional Fu Hsi, or "natural," order. Taking them from right to left, starting at the bottom row, one sees that the hexagrams correspond to those provided by the Fu Hsi chart when read from left to right.

We are now ready to understand why Leibniz, who thought he had invented the binary system in the late 17th century, was so staggered when he first learned of the Fu Hsi sequence from Father Joachim Bouvet, a Jesuit missionary in China. Substitute 0 for each unbroken line, 1 for each broken line, then take the hexagrams in order, reading upward on each, and you get the sequence 000000, 000001, 000010, 000011, . . . , 111111. It is none other than the counting numbers from 0 through 63 expressed in binary notation!

Both Leibniz and Father Bouvet were convinced that Fu Hsi, smitten by divine inspiration, had discovered binary arithmetic, but there is not the slightest evidence for this. The 11th-century *I Ching* scholars had done no more than discover a natural way to arrange the hexagrams. It was not until the time of Leibniz that the Fu Hsi sequence was recognized as being isomorphic with a useful arithmetical notation.

Since the powers of 2 turn up everywhere in mathematical and physical structures, it is not surprising that Chinese scholars have been able to apply the 64 hexagrams to almost everything, from crystal structures to the solar system and the cosmos. Z. D. Sung, in his amusing little book *The Symbols of Yi King* (see the bibliography), tells how he was rotating a matchbox in his hand one day (to simulate the earth's rotation as it goes around the sun) when he suddenly perceived a natural way to generate the eight trigrams at the corners of a cube.

Let the three Cartesian coordinates of a unit cube, x, y, z, indicate the first, second and third digits of a three-digit binary number. Label the corner where the coordinates originate with 000. The other corners are labeled with three-digit binary numbers for 0 through 7, with 0 and 1 indicating the distance of the corner from the origin in each coordinate direction. The eight numbers correspond, of course, to the eight trigrams, with complementary trigrams at diamet-

Figure 150 The Fu Hsi sequence that corresponds to binary numbers 0 through 63

rically opposite corners of the cube *[see Figure 151]*. By a similar procedure, corners of unit hypercubes generate the higher-order polygrams. The 64 hexagrams correspond to six-digit binary numbers at the corners of a six-dimensional hypercube.

Instead of plunging into higher dimensions, Sung divides the cube into 64 smaller cubes that he identifies with the 64 "moods" of the classical syllogism. (The major premise, the minor premise and the conclusion of a syllogism can

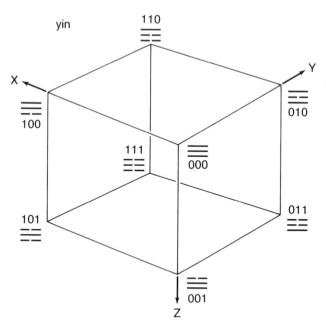

Figure 151   Trigrams generated by a cube

each be of four different forms, giving 64 possible moods.) Sung was probably unaware that this had been done earlier by C. Howard Hinton in his 1904 book *The Fourth Dimension,* London: Allen and Unwin (pages 90–106). Hinton takes a curious step into hyperspace. By considering the four "figures" of each syllogism (an ancient division based on the ordering of the subject, predicate and middle terms), he obtains 256 varieties that he identifies with the 256 cells of a 4-by-4-by-4-by-4 hypercube. Cells corresponding to traditionally valid syllogisms are colored black; the hypercube is then projected onto an ordinary 4-by-4-by-4 cube. The black cells are seen to be symmetrically disposed around one corner of the cube except for one cell that should be black but is not. This led Hinton to the discovery that the anomalous syllogism is valid after all if one applies a more liberal interpretation to syllogisms, one in which the predicate is quantified as well as the subject.

But we have strayed from the *I Ching.* The book (aside from its "Ten Wings," which are appendixes by Confucian metaphysicians) consists essentially of the 64 hexagrams, each followed by a brief explanation of the symbol and six "appended judgments." If the book is to be used as an oracle, one of the hexagrams must be randomly selected, and this must be done in such a way that the rules tell how to transform the chosen hexagram to a second one.

The oldest selection procedure, still followed by those who take the *I Ching* the most seriously, calls for 50 yarrow stalks, each one to two feet long. If yarrow stalks are not obtainable, 50 thin wooden sticks will serve. They should be kept in a lidded receptacle at a spot not lower than a man's shoulders. The *I Ching*, carefully wrapped in clean silk, is kept alongside the sticks.

The book must never be consulted lightly. If you ask it something frivolous or in a skeptical mood, the book gives frivolous or meaningless answers. One should be completely relaxed, physically and mentally. It is essential to think of nothing, throughout the ceremony, other than the question being asked.

Let us assume you are asking the *I Ching* a question and also casting the sticks. Your first step is to unwrap the book, spread the silk on a table and place the book on top. (The cloth protects the *I Ching* from impure surfaces.) An incense burner and the receptacle containing the sticks are placed beside the book. With your back to the south, make three kowtows, touching your forehead to the ground; then, still kneeling, pass the 50 sticks three times through the incense smoke by holding them horizontally and moving your hand in a clockwise circle. Return one stick to its container. It plays no further role in the ceremony.

Put the 49 sticks on the cloth, then with your right hand quickly divide them randomly into two piles. Call the left pile *A*, the other *B*. Take a stick from *B* and put it between the last two fingers of your left hand. With your right hand, push away four sticks at a time from pile *A* until one, two, three, or four sticks remain. Place those sticks between second and third fingers of your left hand. Next diminish pile *B* by pushing away four sticks at a time until one, two, three or four sticks remain. Place these between your left first and second fingers. (This last step can be shortened. Because the sum of the two remainders must be 0, modulo 4, the second remainder is easily calculated from the first.) Your left hand now holds either five or nine sticks. (The possible combinations are 1, 1, 3; 1, 2, 2; 1, 3, 1; and 1, 4, 4.) Put all these sticks to one side.

The remaining sticks are bunched together, and exactly the same dividing procedure is repeated with them, beginning with the random division into two piles. At the finish your left hand will hold either four or eight sticks. (The possible combinations are 1, 1, 2; 1, 2, 1; 1, 3, 4; and 1, 4, 3.) Place them aside, next to the group put aside previously.

Bunch the remaining sticks and repeat the dividing procedure a third time. Your left hand again will hold either four or eight sticks. Put them aside, next to the two groups already there.

The number of sticks that now remain will be either 24, 28, 32 or 36. Count them by groups of four (that is, divide the total number by 4). The quotient will be 6, 7, 8 or 9. These four digits are the ritual numbers, which indicate the character of the bottom line of the hexagram. If the digit is even (6 or 8), the line is yin (broken); if it is odd (7 or 9), the line is yang (unbroken). But the ritual numbers tell you more. Seven and 8 mean that the line (whether yin or yang) is a stable line that cannot be altered. Six and 9 indicate a "moving" line that can be changed (for reasons soon to be explained) to its opposite.

All 49 sticks are now bunched together, and the entire ritual is repeated to obtain the hexagram's second line from the bottom. Four more repetitions give the remaining four lines. The entire ceremony, performed without haste, takes about 20 minutes.

Look up the chosen hexagram in the *I Ching* and study its accompanying text carefully. The text will answer your question and give counsel with reference to the present situation. If all six lines of the hexagram are stable, that is the end of the matter. But if one line or more are moving, change them to their opposites and look up the new hexagram. The commentary will pertain to what you can expect in the future if you follow the counsel of the first hexagram.

After the one or two hexagrams have been written down and the relevant passages in the *I Ching* have been read and meditated on, light another stick of incense, make three more kowtows of gratitude, put the sticks back in their box, rewrap the *I Ching* in its silk and put book and sticks back in their usual high place.

Those too lazy to go through the ancient stick ritual can use a simpler method of casting that has been popular in China for several centuries. It calls for three identical coins, preferably old Chinese coins with square holes. They should be kept polished and should never be removed from their container except when the *I Ching* is being consulted. Observe the same beginning ritual followed for the sticks: kowtowing, kneeling, passing the coins through the incense and so on. Shake the coins in your cupped hands and let them drop simultaneously to the cloth. Having previously decided on which sides of the coins are yin and which yang, consult the following chart to determine whether the throw gives you 6, 7, 8 or 9.

Three yins = 6 (a moving yin line)
Two yins, one yang = 7 (static yang line)
Two yangs, one yin = 8 (static yin line)
Three yangs = 9 (moving yang line)

(If one thinks of the yin side as 2 and the yang side as 3, the sum of the three values will be the desired ritual number.)

Working out the probabilities provided by the stick and coin procedures reveals a subtle difference between the two devination methods. As far as picking the initial hexagram is concerned, the methods are virtually the same, but the probabilities are not the same in choosing the second hexagram. It is not hard to show that in both procedures the probability of choosing a broken line at each of the six steps is 1/2, the same as that of choosing an unbroken line. (This assumes that each time the sticks are randomly divided into piles A and B and A is reduced to one, two, three or four sticks, the probabilities for each of the four outcomes are equal. This is not strictly true, but the deviations from equality are so slight that they have a negligible effect on the final results.) Thus, any hexagram has the same probability of being selected as any other. The two procedures are also alike in giving a probability of 1/4 that a given line will be moving. Since there are six lines, 6/4, or 1½, lines of a hexagram, on the average, will be moving.

When coins are used, the probability that a broken line will change is the same (1/4) as the probability that an unbroken line will change, and similarly the probability is 3/4 that each type of line will remain stable. But when sticks are used, this is not the case. The probability that a broken line will change is 1/16 as compared to 3/16 for an unbroken line (or respective probabilities of 7/16 versus 5/16 that the lines will remain stable). In other words, when sticks are cast, it is three times more likely that an unbroken line will change than a broken one. It is true that any hexagram is as likely to be chosen first as any other, but the more broken lines a hexagram has the more likely it is that it will appear as the second hexagram. Purists who object to coin-casting have sound mathematical support. Not only does the stick ritual discourage frivolous consultation, but also its asymmetry produces a more interesting set of probabilities. We shall say nothing about such impious corruptions as the practice of obtaining the ritual numbers from dollar bills, license plates, telephone numbers and so on.

To readers who may wish to experiment with the *I Ching*, my first recommendation is the Richard Wilhelm and C. F. Baynes translation, rendered into English from the German. Two good paperback translations are also available: one by James Legge and one by John Blofeld.

The Wilhelm–Baynes volume includes the famous foreword by Jung in which he explains the oracular power of the *I Ching* by his theory of "synchronicity," a theory defended by Arthur Koestler in his book *The Roots of Coincidence*. According to Jung, the *I Ching*'s predictions, and relevant events that actually happen, are not causally linked in the Western scientific sense. They are "acausally" related in the Eastern metaphysical sense of being parts of a vast cosmic design that lies beyond the reach of science but is partially accessible to the subconscious mind of the person who casts the sticks. The 64 hexagrams and their meanings are Jungian archetypes, deeply engraved on the collective unconscious of humanity.

Tough-minded skeptics who test the *I Ching* realize at once why the book seems to work. The text is so ambiguous that, no matter what hexagrams are selected, it is always possible to interpret them so that they seem to apply to the question. Indeed, the scope for intuitive interpretation is so great that in China before Mao (I do not know how it is today) there was a large class of professional *I Ching* interpreters whose services were available for a fee on street corners, at fairs and in marketplaces. Surely one reason for the popularity of coin-casting was that it maximized the profits of these fortune-tellers by speeding up their readings.

And if the *I Ching*'s predictions fail to materialize? Well, perhaps the text was not correctly interpreted, or maybe you were not in the right frame of mind when you were tossing the sticks or coins. Besides, the future is not completely determined. The *I Ching*, like the stars of astrology, does no more than indicate probable trends.

Tender-minded believers in the occult, who have not yet consulted the *I Ching* and who long for powerful, mysterious magic, are hereby forewarned. This ancient book's advice can be far more shattering psychologically than the advice of any mere astrologer, palmist, crystal gazer or tea-leaf reader.

ADDENDUM

My remarks on how the probability distributions differ between using the yarrow stalks and the three coins were based on the article by F. van der Blij cited in the bibliography. Persi Diaconis, a Stanford University statistician, called my attention to the fact that van der Blij assumes that when the 49 sticks are divided "randomly," each possible division is equally likely. In actual practice this could hardly be the case. Distribution around the middle (24/25 sticks) is likely to be more peaked than distributions near the extremes (1/48).

An advantage of the coin method is that it provides a fairer distribution of choices. On the other hand, if you believe in Jungian synchronicity you might suppose that whatever acausal forces are operating they would be stronger on a hand division of sticks than on the way one flips a coin. There is also the possibility that if a person is thoroughly familiar with the *I Ching* his unconscious mind may guide the division of sticks toward appropriate passages. I hasten to add that I buy none of the above.

There is an easy way to test the hypothesis that the seemingly beautiful and relevant responses of the I Ching are no more than the result of your mind seizing on ways to apply ambiguous passages to your situation at the moment, like the way devotees of astrology deceive themselves into thinking there are paranormal correlations between their lives and horoscope readings. The test is this. Go carefully through the stick ritual for a friend who is a passionate believer in the powers of the I Ching. But instead of reading the passage selected by the ritual, read instead a passage you selected before you began the ritual. It is highly probable that your friend will be enormously impressed by the I Ching's advice.

It is advisable not to explain the hoax unless you are prepared for an angry reaction. If you yourself are a true believer, you can, of course, argue that precognition and synchronicity guided your choice of a passage. Even if you know the friend's question in advance and search for the most *inappropriate* passage you can find, you'll discover there are always ways of making such a passage appropriate. This is the great secret of the I Ching's success down through the centuries.

There are few signs that the occult revolution is abating in the United States. A dozen or more worthless paperbacks about the I Ching are currently in print, published by firms eager to squeeze as much money as possible from gullible readers who hunger for the paranormal.

BIBLIOGRAPHY

Translations of the I Ching

The I Ching. Translated by James Legge. Dover, 1963. Reprint of the 1899 edition.

The I Ching or Book of Changes, third revised edition. Translated by Richard Wilhelm and Cary F. Baynes. Princeton University Press, 1967.

I Ching: The Book of Change. Translated by John Blofeld. Allen and Unwin, 1965.

About the I Ching

"The Book of Changes." Arthur Waley in The Bulletin of the Museum of Far Eastern Antiquities, Vol. 5, 1933, pages 121–142.

The Symbols of Yi King. Z. D. Sung. China Modern Education, 1934.

"The System of the Book of Changes." Joseph Needham in Science and Civilization in Ancient China, Vol. 2. Cambridge University Press, 1956.

Change: Eight Lectures on the I Ching. Hellmut Wilhelm, Princeton University Press, 1960.

"Combinatorial Aspects of the Hexagrams in the Chinese Book of Changes." F. van der Blij in *Scripta Mathematica,* Vol. 28, 1967, pages 37 – 49.

Jung, Synchronicity, and Human Destiny. Ira Progoff. Julian, 1973. See Chapter 3.

The I Ching and Modern Man. Jung Young Lee. University Books, 1975.

"A Reordering of the Hexagrams of the *I Ching.*" Stephen E. McKenna and Victor H. Mair in *Philosophy East and West,* Vol. 29, 1979, pages 421 – 441.

Lectures on the I Ching: Constancy and Change. Richard Wilhelm. Princeton University Press, 1979.

Researches on the I Ching. Julian Konstantinovich Shchutski. Princeton University Press, 1980.

CHAPTER TWENTY-ONE

The Laffer Curve

The Kettle-Griffith-Moynihan Scheme for a New Electricity Supply,
Traveling in the Olden Times,[4] American Lake Poetry, the Strangest Dream
that was ever Halfdreamt.[5]

[4]I've lost the place, where was I?

[5]Something happened that time I was asleep, torn letters or was there snow?

—JAMES JOYCE, *Finnegans Wake*

Economists love to draw curves. In the early decades of modern capitalism, classical economists were fond of explaining prices by constructing supply and demand graphs such as the one shown in Figure 152. If the price of a commodity is on the level indicated by the broken line *a*, it is easy to see from where this line crosses the curves that people will buy less of the product. Since the seller will have an oversupply, he will lower its price to get rid of it. If the prices are on the lower level of the broken line *b*, increased demand will bid up the product's price and the seller will produce more.

These up and down forces stabilize the price at *E*, the equilibrium point where the amounts demanded and supplied are equal. At this point, according to early classical theory, the seller maximizes profit. If there is a general increase in demand, with supply constant, the demand curve shifts to the right and *E* rises. If there is a general increase in supply, with demand constant, the supply curve shifts to the right and *E* falls. If both curves move to the left or the right the same distance, *E* stays at the same level.

These curves are still indispensable because supply and demand play basic roles in any economy, even one without free markets; but these days economists refer to them less, because in a mixed economy such as ours hundreds of

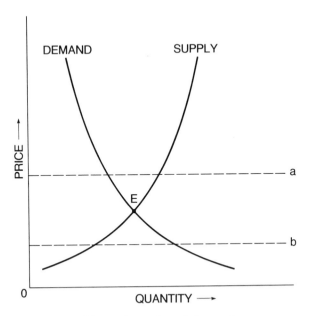

Figure 152   Classical supply and demand curves

variables play havoc with the curves. The government, by innumerable strata-
gems, keeps many prices far above or below what they would be in a free
market. Organized labor pushes up wages, and companies pass the increases
along to prices, in what Arthur M. Okun of the Brookings Institutions calls "the
invisible handshake." Oligopolists find subtle ways of getting together to avoid
market fluctuations, something they must do to remain efficient.

In the 1960's, when Keynesian economics was still carrying all before it
("We are all Keynesians now," said Richard Nixon), many economists were
impressed by the Phillips curve. This curve was first proposed in 1958 by the
London economist Alban William Housego Phillips and applied to the U.S.
economy in 1960 by neoKeynesians Paul A. Samuelson and Robert M. Solow.
As you can see in Figure 153, a typical Phillips curve plots the inverse relation
between unemployment and inflation. By taking into account the ability of
labor and business to administer prices, the Phillips curve indicates that the
double goals of full employment and price stability are not compatible in a
mixed economy. Full employment (F) is attainable only at the cost of steady
inflation. Stable prices (zero inflation) are impossible without high unemployment (U).

What to do? The best we can hope for, implies the curve, is to find a
reasonable trade-off that does the minimum amount of harm. If prices rise too
high, let a recession pull them down. If too many people are out of work, let

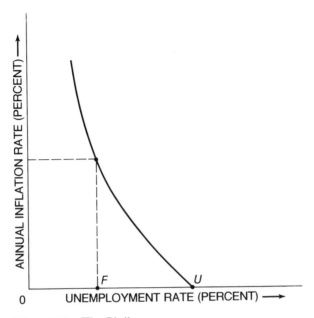

Figure 153   The Phillips curve

inflation restore their jobs. With luck a government may find a point on the curve where "normal" unemployment will combine with an acceptable mild inflation of, say, 4 or 5 percent per year.

While economists were arguing about the "cruel dilemma" posed by the Phillips curve — the difficulty of finding a trade-off that would not lead to either a deep recession or a galloping inflation — a funny thing happened. During the late 1950's and early 1960's the economy got itself into the mysterious state of "stagflation" where, contrary to the curve, unemployment and inflation began to rise simultaneously. The Phillips curve started to disintegrate.

Keynesians struggled to rescue the curve. It was soon obvious that there is no such thing as a Phillips curve that is stable in the short run. The curve can be drawn dozens of ways, depending on what variables (including psychological expectations) are taken into account, and it varies widely from time to time and place to place. Is there a Phillips curve that is stable in the long run? Some say yes, some say no. Even if there is, economists disagree on how to apply it. Should the government try somehow to slide up and down the curve, with inflation and unemployment fluctuating like a seesaw? Should it try "looping" around the curve in various risky ways?

According to Keynesians, a force called demand-pull tries to twist the curve into a vertical straight line, while another force called cost-push tries to twist it into a horizontal line. The long-run curve compromises with a steep downward slope. What is needed, of course, is some way of shifting the entire curve back down and to the left to allow trade-offs that will not lead to social chaos. Some economists, for example John Kenneth Galbraith, believe this can be done only by combining fiscal and monetary policies with wage and price controls. Nothing could be worse says Milton Friedman. In Friedman's monetarist view the long-run Phillips curve is a vertical line at the "natural rate" of unemployment, and any tradeoff effort to reduce unemployment below that line will set off an explosive inflation.

The Phillips curve, Daniel Bell wrote in 1980 (summarizing earlier remarks by Solow), "provided more employment for economists . . . than any public-works program since the construction of the Erie Canal." If unemployment is plotted against inflation for the 1960's, the result is a reasonably smooth curve. But if the same chronological plotting is done for the 1970's, as the U.S. economy drifted deeper into stagflation, the result is what the Wonnacotts, in their textbook *Economics,* call a "mess." Today the Phillips curve has become little more than an out-of-focus symbol of the fact that inflation and unemployment are not independent evils but are functionally linked in complex ways that nobody is yet able to understand.

Now, as a result of the upsurge of interest in "supply-side" economics, the curve of the hour is a brand-new one called, with strangely resonant overtones, the Laffer curve. Arthur B. Laffer is a 41-year-old professor of business at the University of Southern California. The curve was named and first publicized by Jude Wanniski, a former writer for *The Wall Street Journal,* in his bible of supply-side theory, confidently titled *The Way the World Works.* Figure 154 shows how Wanniski orients the Laffer curve at the beginning of his Chapter 6.

Is it not a thing of beauty bare? As any child can see from inspecting the curve's lower end, if the government drops its tax rate to nothing it gets nothing. And if it raises its tax rate to 100 percent, it also gets nothing. Why? Because in that case nobody will work for wages. If all income went to the state, people would revert to a barter economy in which a painter paints a dentist's house only if the dentist caps one of the painter's teeth.

The Laffer curve gets more interesting when we slide along its arm toward the center. At point A, where taxes are not quite 100 percent, people will find it to their benefit to take some of their income in taxable wages. At point B the economy hums along with unfettered high production, but because tax rates are low the government gets the same small amount it would get if taxes were at A.

Now look at point E at the extreme right of the curve. That is where the tax rate maximizes government revenue. If taxes fall below E, that may stimulate production, but it obviously diminishes government revenue. Because E, by definition, is the point of maximum revenue, the government also must get less

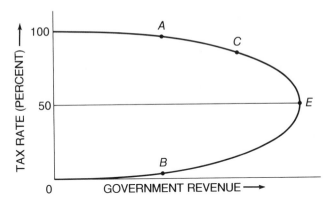

Figure 154   The Laffer curve

if taxes rise above *E*. The supply-siders stress many reasons for this being so. Some rich people find it unprofitable to work as productively as before. Some escape from excessive tax burdens by finding unproductive "shelters." Some even move to another country where taxes are low. If the government is relying on high taxes for welfare programs, millions of people are encouraged not to work at all. Why work if you can get almost the same income from welfare? Big corporations spend less on research and development. Entrepreneurs, the backbone of dynamic growth, are less willing to take risks. As a result of these factors and others, the economy becomes sluggish and tax revenues decline.

It is important to understand, Wanniski tells us, that *E* is not necessarily at the 50 percent level, although it could be. The shape of the Laffer curve obviously changes with circumstances. Thus, in time of war, when people and business are persuaded that a sacrificial effort is essential, they are willing to accept a high tax rate while they keep production booming. In peacetime they are less altruistic.

Now, the heart of the supply-side argument is the conviction that our current economy is somewhere near *C*, far too high on the Laffer curve. Lowering taxes (which some supply-siders believe calls for huge cuts in welfare spending) will give the supply side of the economy such a shot in the arm that the U.S. will slide down the Laffer curve to point *E*, perhaps not right away but soon. Tax revenues eventually will rise enough to take care of increased funding of the military, stagflation will end, dynamic growth will begin, the budget will be balanced by 1984 and the American dream will regain its luster.

Of course, supply and demand are always intertwined, but the supply-siders call themselves supply-siders in order to emphasize how they differ from neo-Keynesians. John Maynard Keynes stressed the importance of maintaining demand by minimum-wage laws and welfare payments. The Lafferites turn this around and stress the importance of stimulating supply. With the government off the back of business production will soar, new inventions will be made, more people will be employed and real wages will rise. Everyone benefits, particularly the poor, as prosperity trickles down from the heights.

The second book to gild the virtues of Lafferism is George Gilder's *Wealth and Poverty*. The title intentionally plays on the title of Henry George's best seller *Progress and Poverty*, which created a stir late in the 19th century by recommending the abolition of all taxes except a single tax on land. Gilder's book is more impassioned than Wanniski's. "Regressive taxes help the poor!" Gilder exclaims on page 188. William Safire once described capitalism as the "good that can come from greed." Gilder is furious when people talk like that; he finds capitalism motivated by the good that comes from "giving." By this he means that the best way to give the poor what they want, particularly the unemployed young men and women of minority groups, is to leave the free market alone so that the economy will start growing again.

The trouble with the Laffer curve is that, like the Phillips curve, it is too simple to be of any service except as the symbol of a concept. In the case of the Laffer curve the concept is both ancient and trivially true — namely that when taxes are too high they are counterproductive. The problem is how to define "too high." No economist has the foggiest notion of what a Laffer curve really looks like except in the neighborhood of its end points. Even if economists did know, they would not know where to put the economy on it. Neoconservative Irving Kristol, defending supply-side economics in *Commentary*, writes that he cannot say where we are on the Laffer curve, but he is sure we are "too far up." President Reagan's across-the-board tax cuts are, he says, just what we need in order to slide the economy toward point E.

To bring Laffer's curve more into line with the complexities of a mixed economy dominated by what Galbraith likes to call the "technostructure," and also with other variables that distort the curve, I have devised what I call the neo-Laffer (NL) curve. The NL curve is shown in Figure 155. Observe that near its end points this lovely curve closely resembles the old Laffer curve, proving that it was not a totally worthless first approximation. As the curve moves into the complexities of the real world, however, it enters what I call the "technosnarl." In this region I have based the curve on a sophisticated statistical analysis (provided by Persi Diaconis, a statistician at Stanford University) of the best available data for the U.S. economy over the past 50 years. Since the data are represented on the graph by a swarm of densely packed points, the actual shape of the curve is somwhat arbitrary. Nevertheless, it dramatizes a number of significant insights.

Consider any value r on the revenue axis within the segment directly below

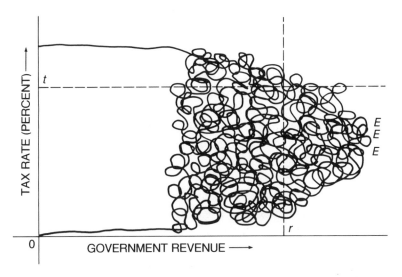

Figure 155   The neo-Laffer (NL) curve

the technosnarl. A vertical line through *r* intersects the snarl at multiple points. These points represent values on the tax-rate axis that are most likely to produce revenue *r*. Note that this also applies to the maximum value of *r*, producing multiple points *E* on the technosnarl. In brief, more than one tax rate can maximize government revenue.

Consider any value *t* on the tax-rate axis within the segment directly to the left of the technosnarl. A horizontal line through *t* also intersects the snarl at multiple points. These points represent values on the revenue axis that are most likely to result from tax rate *t*.

Note that at some intersection points lowering taxes from a given tax rate will lower revenue, and that at other points for the same tax rate it will raise revenue. Even if we could determine at which point to put the economy, it is not clear from the snarl just what fiscal and monetary policies would move the economy fastest along the curve to the nearest point *E*.

Like the old Laffer curve, the new one is also metaphorical, though clearly a better model of the real world. Since it is a statistical reflection of human behavior, its shape constantly changes, like the Phillips curve, in unpredictable ways. Hence the curve is best represented by a motion picture that captures its protean character. Because it takes so long to gather data and even longer to analyze all the shift parameters, by the time an NL curve is drawn it is out of date and not very useful. I have been told in confidence, however, by one of Jack Anderson's more reliable informants that the Smith Richardson Foundation has secretly funded a multimillion-dollar project at Stanford Research

International to study ways of improving the construction of NL curves. It is possible that with better software, using the fast Cray computer at the Lawrence Livermore Laboratory, one will be able to assign current probability values to the multiple intersection points. If one can do so, the NL curve could become a valuable forecasting tool for rational Federal decisions.

The Lafferites combine supreme self-confidence with a supremely low opinion of their detractors. Of the 18 economists who have won Nobel prizes, only two, Milton Friedman and Gunnar Myrdal, appear in the index of *The Way the World Works*. Not even Alan Greenspan, now of the abandoned "old right," gets a mention. You might suppose that, since Friedman and Wanniski are both mentors to conservatives, Wanniski would have a high opinion of Friedman. Not so. Wanniski goes to great lengths in his book to explain why three famous economic models — Marxian, Keynesian and Friedmanian — are all wrong. They cannot even explain why the economy crashed in 1929.

There is now an enormous literature on the many causes of the crash, much of it written by eminent economists. We can throw it all away. Wanniski has figured out the real reason. There would have been nothing wrong with the stock market if Herbert Hoover had just left it alone. Instead he and Congress made a stupid political blunder. Writes Wanniski: "The stock market Crash of 1929 and the Great Depression ensued because of the passage of the Smoot-Hawley Tariff Act of 1930."

How did it happen that the crash occurred in October of the previous year? It is simple. The stock market, says Wanniski, anticipated the dire consequences of the coming restraints on free trade. Not all supply-siders agree. Jack Kemp, the New York congressman who coauthored the Kemp–Roth tax bill (which paved the way for Reagan's fiscal program), is one who does. In Kemp's rousing book *An American Renaissance,* he assures us that Wanniski has "demonstrated beyond any reasonable doubt" the truth of his remarkable discovery.

What do professional economists make of radical supply-side theory? Most of them, including the most conservative, regard it in much the same way as astronomers regard the theories of Immanuel Velikovsky. To Galbraith it is "a relatively sophisticated form of fraud." Walter W. Heller has likened it to laetrile, and Solow terms it "snake oil." Vice-president Bush has called it "voodoo economics." Herbert Stein labeled it "punk economics" (as in "punk rock"), and Martin Feldstein described it as "excess rhetorical baggage." Nevertheless, the books by Wanniski, Gilder and Kemp are said to have much influence in the current Administration.

Lafferites enjoy heaping praise on one another. Laffer, the hero of Wanniski's book, is quoted on the back of the paperbound edition as saying: "In all honesty, I believe it is the best book on economics ever written." Kristol, on the front cover, is more restrained. He thinks it is "the best economic primer since Adam Smith." Gilder asserts Wanniski "has achieved an overnight influence of nearly Keynesian proportions." Gilder has been greeted with similar euphoria. David Stockman, President Reagan's budget director, has hailed *Wealth and Poverty* as "Promethean in its intellectual power and insight. It shatters once and for all the Keynesian and welfare-state illusions that burden the failed conventional wisdom of our era."

How puzzled the President must be by the violent clash between his old friend Friedman and his Lafferite advisers! (The clash is not only over Friedman's monetary views but also over his distaste for the supply-side "gold bugs" who are urging an immediate return to the gold standard.) In the business section of *The New York Times,* Wanniski's attack on Friedman was vitriolic. The burden of it is that although Friedman is "barely five feet tall," he "weighs" so much that he is now an enormous "deadweight burden" on the backs of Menachem Begin, Margaret Thatcher, Ronald Reagan and the U.S. economy.

Will the Lafferism of the Administration succeed, or will it, as many economists fear, eventually plunge the nation into higher inflation and higher unemployment? Economists cannot know. The technosnarl is too snarly. The idle rich might not invest their tax savings, as Lafferites predict, but might spend it on increased conspicuous consumption. The hardworking poor and middle class might decide to work less productively, not more. Big corporations and conglomerates might do little with their tax savings except acquire other companies.

Of course, ideologues of all persuasions think they know exactly how the economy will respond to the Administration's strange mixture of Lafferism and monetarism. Indeed, their self-confidence is so vast, and their ability to rationalize so crafty, that one cannot imagine any scenario for the next few years, that they would regard as falsifying their dogma. The failure of any prediction can always be blamed on quirky political decisions or unforeseen historical events. It is inconceivable, for example, that Friedman would consider the triple-digit inflation in Israel or the recent riots in Britain or high U.S. stagflation in 1983 as suggesting the slightest blemish on his monetarist views even though he enthusiastically supported Begin, Thatcher and Reagan, and all three have in turn been strongly influenced by Friedman's brand of monetarism.

As for the Lafferites, they have all kinds of outs in case Reagan's policies lead to disaster. Some will blame it on Friedman. Others may follow an escape plan mapped out by William F. Buckley. Although the Administration's tax and budget cuts have been called the biggest in American history, Buckley thinks both cuts are not big enough. "The trouble with the Reagan tax cuts," he wrote in *National Review* (July 24), "is (a) they are insufficient, and insufficiently targeted; and (b) the cuts in the budget are equally insufficient. . . . You cannot make long-range, significant cuts by concentrating on only a single one-third of the budget. It is the equivalent of saying you are going to lose weight by exercising only your right leg."

One can hope that President Reagan will not try to reconcile these conflicting conservative views by resorting to astrology. This possibility is not quite as remote as one might think. In an interview with Angela Fox Dunn the President said he followed the daily advice for his sign in the syndicated horoscope of Carroll Righter. Born on February 6, Reagan is an Aquarian. "I believe you'll find," he told Dunn, "that 80 percent of the people in New York's Hall of Fame are Aquarians."

President Reagan and his wife Nancy have for many years been personal friends of both Righter (who advises Gloria Swanson and other Hollywood figures) and the astrologer Jeane Dixon, who lives in Washington. "I'm not considered one of his advisers," Dixon cryptically told newspaper columnist Warren Hinckle, "but I advise him." Joyce Jillson, who writes a syndicated astrology column for *The Chicago Tribune* and has among her clients several Hollywood studios and multinational corporations, says that in 1980 Reagan aides paid her $1,200 for horoscopes on eight prospective vice-presidential candidates. The White House communications director has, however, called her a liar. Michael Kramer writes in *New York Magazine:* "Ronald Reagan, says Ronald Reagan, is a nice, well-intentioned man who loves his family, likes to consult his horoscope before making major decisions, and cries when he watches *Little House on the Prairie.*"

Will the President seek help from the zodiac in trying to decide whether to follow Friedman or Laffer or someone else? One may never know. As the Yale economist William Nordhaus put it (*The New York Times,* August 9, 1981): "We can only hope that supply-side economics turns out to be laetrile rather than thalidomide."

ADDENDUM

Book collections of my *Scientific American* columns have taken them in rough chronological order, but the previous chapter is a rare exception. It was my last column, appearing in the December issue of 1981.

Ronald Wilson Reagan had been president for a year. It may be hard to believe today, but the extreme supply-siders had convinced him that if he lowered taxes it would give the economy such a shove that defense spending could be increased, the entitlement programs preserved, and the budget balanced by 1984 with a healthy surplus. At least that's what Reagan told the voters. There is now some evidence that he and his advisors anticipated a large deficit but kept this secret because they were convinced that only such a deficit would persuade Congress to dismantle welfare. In any case, everybody except a few diehard supply-siders such as Arthur Laffer, Jack Kemp and his writer friends, the economist Paul Craig Roberts and, of course, the president himself now realizes that the 1980 campaign promises were fantasy. The one big achievement of the Reagan administration has been lowering the inflation rate — but at what a cost!

It's no economic mystery that inflation can be checked if you pay the Keynesian price of a severe recession and a monstrous deficit. Indeed, one of the many ironies of Reagan's career is that although he began his first term as a radical rightist — it's not easy to pin the label "conservative" on him — he ended the term by reviving a moribund Keynes. Economists now generally agree that the recession was caused by the Fed's tight money policy (plus other things), and it ended only when the Fed abruptly eased the money supply in 1983, again aided by other factors. These other factors included rising government spending on defense, increased purchasing power created by tax cuts, and the sheer fact that depressions are cyclical. Whatever the multiple reasons, Reagan was lucky. The depression ended just before his 1984 campaign. The point, however, is this: Everything happened according to classical Keynesian doctrine.

"Reagan became the ultimate Keynesian," was how Lester Thurow put it (see Karen Arenson's article, "Heroes of the Economic Recovery," *New York Times*, Sunday, January 19, 1984). "Regardless of what he said he was doing, it was simply the old Keynesian medicine at work, stop and go economics. It got us out of the worst recession since the depression, and we're now in the go phase. But the problem is that we will eventually stop."

At the time I am typing this (late 1985), Reaganism is falling apart, and it is impossible to predict how Congress will eventually handle the deficit catastrophe without cutting defense, chipping away at social security, raising taxes, or some combination of the three. One thing, though, is clear. The Laffer curve is a joke. It seemed like a good idea to reprint my column about it now rather than hold it for the final volume of the series 10 years from now when the curve will be remembered (if at all) only as a quaint curiosity. By then we should know how the nation met the deficit crisis. Unless taxes are steeply raised in the next few years, the only viable alternative short of a major war seems to be a pumping up of the money supply (equivalent almost to printing money) to wipe out debts at the cost of an unthinkable inflation.

The big stumbling blocks are Reagan's persistent hope for an economic miracle, and his declaration that he would never raise taxes or trim defense spending and Social Security. When he was governor of California, he similarly did his best to reduce taxes and chop welfare. "The entire graduated income tax structure was created by Karl Marx," Reagan said in 1966. He wanted to declare war on Vietnam ("We could pave the whole country and put parking stripes on it, and still be home for Christmas"). He described welfare recipients as "a faceless mass waiting for a handout." Yet when the state budget jumped up 122 percent, Reagan approved the largest tax increase in the history of California. By the end of his second term we will know if he is capable of approving a tax increase for the nation even though he promised in 1984 that this would be done "over my dead body." Reagan knows, of course, that the Democrats will never stop recalling Walter Mondale's campaign prediction that the administration would be forced to raise taxes to prevent the deficit from wrecking the economy.

Reagan wants to go down in history as the far-sighted president who reversed what he sees as evil drifts toward socialism, toward accommodation with the godless Soviet Union, and toward a general decay of morality and Christian faith. He wants to be remembered as the David who slew the Goliath of Big Government, unraveled the welfare state and put a free market back at the center of the economy. More likely he will be seen by future historians as another Herbert Hoover. ("We should soon, with the help of God, be in sight of the day when poverty will be banished from the nation," said Hoover just before the Big Crash of 1929). Conservative writer William Safire (in an August 1985 column) described Reagan's present strategy as the "masterly inactivity" of a leader too stubborn to go back on his "demogogic pre-election promises," standing on the bridge of the ship of state "smiling into the fog, as we head toward his trillion-dollar iceberg."

My neo-Laffer curve produced a flood of letters, many from indignant readers who actually took the curve seriously. Some conservative economists congratulated me for saying what they had avoided saying out of respect for a newly elected, enormously popular president. David Warsh, of the *Boston Globe*, wrote a feature article about my curve ("No Laffing in D.C. This Week," December 15, 1981). I was taken to task by several economists for implying that the Laffer curve was something the profession considered significant, when in fact it was the product of media hype. Three letters, two blasting me and one defending me, were published in *Scientific American* (March, 1982).

Neither supply-siders nor monetarists buy the old Phillips curve. Indeed, as we have seen, they apparently sold Reagan on the dream that inflation could be checked without either wage and price controls or a recession. In some cases their understanding of economic realities behind the curve was on a primitive level. Leonard Silk, describing the 1982 economic summit conference at Versailles (*New York Times,* June 11, 1982) reported a briefing of reporters by supply-sider Donald Regan, then Secretary of the Treasury. "If you recall the Phillips curve," said Regan, "that's where the more you have of inflation, the more unemployment you'll have — and the less inflation, the less unemployment."

Of course, Regan had it exactly wrong. The curve shows inflation and unemployment moving in *opposite* directions. Regan, Silk reminded his readers, had been an English major at Harvard. There was sharp disagreement at the conference over whether the Phillips curve could be made to go away, though everybody agreed it would be great if it would. Meanwhile, added Silk, "The president is still clinging to the Laffer curve, with its claimed relationship between lower taxes, and higher production, national income, and tax revenues."

Jude Wanniski has been almost as subdued lately as Milton Friedman, but Laffer and Gilder continue to bubble over with supply-side enthusiasm. In the *Washington Post* (August 20, 1985) Laffer maintained that the drop in inflation was not the result of tight money; it resulted from an increase in the supply of goods. "The size of our national debt is not a crisis situation," he said a few days later (*Sacramento Bee*, August 29, 1985). "It is by no means a reason to overturn Reaganomics. It's far better to keep tax rates low and run temporary deficits than it is to raise tax rates and destroy economic growth." Congress, according to Laffer, is solely to blame for the big deficit because it refused to cut nondefense spending enough.

Gilder's latest work, *The Spirit of Enterprise* (1984) is another breathless hymn to the invisible hand of Adam Smith. "No one understands the entrepreneurial spirit and the entrepreneurial basis of economic growth better than George Gilder," said his friend Irving Kristol, who seems unable to curb hyperbole whenever he is asked to supply a jacket blurb for a book by a supply-sider.

Keynesian economists naturally see Gilder's book in a different light. "Only someone with a sense of humor could survive reading this book," commented Robert Solow (*New Republic,* October 22, 1984). "And no one with any trace of a sense of humor could have written it. . . . the prose is mind numbing. . . . if he wrote a chapter in praise of lettuce, it could turn you against green leafy vegetables forever."

Here is how Solow said what I tried to convey with my neo-Laffer curve:

> The truth is that the tradeoff between incentive and equity in taxation is a complicated and tough issue of public policy. Our tax system probably does a terrible job. It is riddled with loopholes that tend to direct energy into unproductive activities. It does not achieve much real equity (and even less nowadays). If we had the will to reform it — which we do not — we

might gain on both the equity and incentive sides. But it is neither clever nor honorable to do as Gilder does and submerge the equity issues in undocumented claims and claptrap. It irresistibly reminds one of what Bernard Shaw is supposed to have said to Samuel Goldwyn as they negotiated over the royalties for a film version of one of Shaw's plays: "Mr. Goldwyn, you seem to be interested only in art, while I care only about money."

David Stockman, as we all know, resigned in 1985 as budget director after another outburst of harsh and honest words about Reaganomics. Incidentally, it has been reported that Stockman used to baby-sit for Daniel Patrick Moynihan. Contrary to James Joyce's prophecy (see the epigraph of this chapter), Moynihan never bought his baby-sitter's supply-side mythology. Exactly where Stockman stands today on the "promethean" intellects of Wanniski and Gilder is not clear. Perhaps he will tell us in the memoirs he is writing, and for which Harper and Row paid him more than $2 million. To quote Joyce again: "Stockins of Winning's Folly Merryfalls. . . . Godamedy, you're a delville of a tolkar!"

BIBLIOGRAPHY

Supply-Side Defended

The Way the World Works: How Economists Fail and Succeed. Jude Wanniski. Basic Books, 1978.

The Economics of the Tax Revolt. Arthur B. Laffer and Jan P. Seymour. Harcourt Brace Jovanovich, 1979.

An American Renaissance. Jack Kemp. Harper and Row, 1979.

Reaganomics: Supply-Side Economics in Action. Bruce R. Bartlett. Arlington, 1980.

Wealth and Poverty, George Gilder. Basic Books, 1981.

The Reagan Revolution. Rowland Evans and Robert Novak. Dutton, 1981.

"Ideology and Supply-Side Economics." Irving Kristol in *Commentary,* April, 1981, pages 48–54.

"The Burden of Friedman's Monetarism." Jude Wanniski in *The New York Times,* Sunday, July 26, 1981.

"No Shrinking Supply-Sider: Economist Arthur Laffer Keeps the Faith." Kathryn M. Welling in *Barron's,* December 21, 1981.

The Spirit of Enterprise. George Gilder. Simon and Schuster, 1984.

Supply-side attacked

"Changing Perspectives on the Laffer Curve." Martin Gardner in the *British Journal of Econometric Hogwash*, Vol. 34, 1980, pages 7,316–7,349.

Greed Is Not Enough: Reaganomics. Robert Lekachman. Pantheon, 1982.

The Phillips Curve

"The Relation Between Unemployment and the Rate of Change of Money Wages in the United Kingdom." A. W. Phillips in *Economica*, November, 1958, pages 282–289.

"Unemployment and Inflation: The Cruel Dilemma." James Tobin in *Prices: Issues in Theory, Practice, and Public Policy*, edited by Almarin Phillips and Oliver E. Williamson. University of Pennsylvania Press, 1967.

"Phillips Curves, Expectations of Inflation and Optimal Unemployment Over Time." Edmund S. Phelps in *Economica*, August, 1967, pages 254–281.

"Down the Phillips Curve with Gun and Camera." Robert M. Solow in *Inflation, Trade and Taxes*, edited by David A. Belsley and others. Ohio State University Press, 1976.

"Aggregate Supply: How Can Inflation and Unemployment Coexist?" Paul and Ronald Wonnacott in *Economics*, second edition, McGraw-Hill, 1979.

"What We Know and Don't Know About Inflation." Robert M. Solow in *Technology Review*, Vol. 81, 1979, pages 30–46.

Reagan and Astrology

Where's the Rest of Me? Ronald Reagan and Richard G. Hubler. Karz, 1981. Reprint of 1965 edition.

"Ronald Reagan's Affinity for Stargazers." Warren Hinkle in the *San Francisco Chronicle*, July 19, 1980.

"I Was Always the Last One Chosen: Reagan Talks About His Childhood, His Heroes, His Horoscope." Angela Fox Dunn in the *Washington Post*, July 13, 1980. This interview was distributed by the *Los Angeles Times*, which, curiously, never published it. It ran in many newspapers around the country.

"When Reagan Spoke From the Heart." Michael Kramer in *New York*, July 21, 1980.

"It's In the Stars, Mr. President." Frederic Golden in *Discover*, January, 1988, page 82.

"Horoscopes: Fans Bask in Sun Signs." Penelope McMillan in the *Los Angeles Times*, July 5, 1985.

Index